Youth Fund of National Natural Science Foundation of China (31600437);
Natural Science Foundation of Guangdong Province (2020A1515011072);
"Double first class" Construction Project of South China University of Technology(D6201670);
The Fundamental Research Funds for the Central Universities(D2200820)

Planning for Sea Level Rise and Land Use Change: Adaptive Conservation Planning and Reserve Design to Conserve Biodiversity

Mingjian Zhu

中国建筑工业出版社

图书在版编目（CIP）数据

应对海平面上升和土地利用快速变化的生态保护规划 = Planning for Sea Level Rise and Land Use Change : Adaptive Conservation Planning and Reserve Design to Conserve Biodiversity : 英文 / 祝明建著 . -- 北京 : 中国建筑工业出版社 , 2020.12

ISBN 978-7-112-25505-4

Ⅰ . ①应… Ⅱ . ①祝… Ⅲ . ①海平面变化—影响—生态环境保护—环境保护规划—研究—英文②土地利用—影响—生态环境保护—环境保护规划—研究—英文 Ⅳ . ① X321

中国版本图书馆 CIP 数据核字（2020）第 197224 号

责任编辑：吴　绫
文字编辑：孙　硕
责任校对：张　颖

Planning for Sea Level Rise and Land Use Change:
Adaptive Conservation Planning and Reserve
Design to Conserve Biodiversity

Mingjian Zhu

*

中国建筑工业出版社出版、发行（北京海淀三里河路9号）
各地新华书店、建筑书店经销
北京光大印艺文化发展有限公司制版
北京中科印刷有限公司印刷

*

开本：787毫米×1092毫米　1/16　印张：9¼　字数：228千字
2021年1月第一版　2021年1月第一次印刷
定价：58.00元

ISBN 978-7-112-25505-4
（36499）

TABLE OF CONTENTS

LIST OF TABLES

LIST OF FIGURES

PREFACE

Coupled with human population growth and increasing urbanization, sea level rise is likely to be one of the most important challenges for biodiversity conservation in coastal areas of the world. To facilitate biodiversity adaptation to sea level rise and land use change, designing functional reserve networks is essential for conservation planners and coastal decision-makers.

Associate professor Mingjian Zhu at South China University of Technology is very committed to working in the combination of conservation-based ecological science and design and planning to ensure the conservation of biodiversity and ecosystem services. He has shown particular aptitude in applying Geographic Information Systems (GIS) tools to facilitate conservation assessment and design.

This book introduces a sea level rise impact assessment and adaptation project and develops an integrated modeling framework on the use of conservation reserve design tools, cost efficiency considerations, and the value of high resolution sea level rise data and models for impact assessment. The work is both high quality and contributes to an extremely important emerging field regarding assessing potential coastal impacts of climate change and making science-based recommendations for avoiding, minimizing, mitigating, and adapting to the impacts of sea level rise. Methodologies developed in this book are transferable to other low-lying coastal areas in the world for sea level rise adaptation.

This book has unique value in ecological planning and design especially conservation design under climate change. As an encouragement and support for this important interdisciplinary research field of conservation biology, urban and rural planning and landscape architecture, I would like to preface this book and recommend it to students, faculty and professionals in relevant disciplines.

Professor
Deputy Director of Landscape Architecture Department
College of Architecture and Urban Planning
Tongji University
Nov. 26, 2020

FOREWORD

Coupled with human population growth and increasing urbanization, sea level rise (SLR) is likely to be one of the most important challenges for biodiversity conservation in Florida. In response to SLR and future land use change, designing functional reserve networks is essential to facilitate biodiversity adaptation to SLR. The aim of this dissertation is to develop and assess an integrated SLR adaptive conservation planning model that incorporates economic concepts in the planning process with different model-dataset combinations in the Matanzas River Basin of Northeast Florida.

Three issues are discussed in three chapters. In Chapter 2,I develop an integrated modeling process using a geomorphological model (SLAMM), species habitat models, and conservation prioritization (Zonation) to identify conservation priorities in response to SLR. I compare the SLR adaptive conservation priorities with existing Florida conservation layersincluding current reserves, the Florida Ecological Greenways Network (FEGN) and the Critical Lands and Waters Identification Project (CLIP). Results show that current reserves are not adequate to protect some of the most important conservation priorities as SLR occurs but that the updated FEGN and CLIP do serve as a good foundation for future conservation decisions.

Chapter 3 explores how economic concepts can be integrated into SLR adaptive conservation planning. The economic side of conservation analysis in this research demonstrates that conservation priorities with consideration of costs could be significantly different from conservation prioritization without consideration of costs. Understanding the economic side of conservation planning will help other coastal resource managers make informed decisions about where/how to allocate resources more wisely to facilitate biodiversity adaptation to SLR.

Chapter 4 investigates the value of information (data accuracy & model sophistication) in SLR adaptive conservation planning. The results suggest that the combination of process-based coastal impact modeling and high-resolution elevation data could considerably improve the identification of potential species habitat and conservation priorities that are adaptive to SLR. The less accurate model and elevation dataset when used in SLR adaptive conservation designscould fail to identify areas that have high conservation value.

Chapter 5 includes recommendations for land protection and growth management to facilitate biodiversity adaptation to SLR.

CHAPTER 1
INTRODUCTION

1.1 Challenges for Ecological Conservation in Florida

1.1.1 Climate Change and Sea Level Rise

The global annual average temperature has risen approximately 0.7°C over the 20th century (IPCC, 2013). Climate change has profound implications for biodiversity and ecosystem conservation (Lawler, 2009).The effects of climate change on the earth's natural systems have been discussed for several decades in conservation biology (Peters & Darling, 1985; Noss, 2011). Climate change could lead to increases in the frequency and severity of wildfires (Westerling et al., 2006; Lawler, 2009), intensification of storms and hurricanes (Bengtsson et al., 2009), and rising sea levels (IPCC, 2013). Recent studies show that climate change could soon become one of the major causes of the global species extinction (Sala et al., 2000; Heller & Zavaleta, 2009; Noss, 2011). Evidence of the effects of climate change on species and natural communities have already been demonstrated as many species are migrating poleward and upslope in response to a warmer climate (Root et al., 2003; Noss, 2011). In addition, many species are advancing their spring phenology since warmer temperatures are often coming earlier in spring than in the past (Root et al., 2003; Parmesan & Yohe, 2003; Noss, 2011).

Among the effects of climate change, sea level rise (SLR) is considered as the most certain, immediate, widespread and visible result of climate change (Pilkey & Yong, 2009; Noss, 2011). Based on records from the1990s to 2010, average sea level increased 3.26 mm per year globally (Figure 1-1). The newly released 5th assessment report (AR5) from the Intergovernmental Panel on Climate Change (IPCC) estimates a potential SLR of 40 to 60 cm by the late 21st century and 1.0 m as the worst case scenario by 2100 (Church et al.,

2013). However, the report also concluded that "sea levels could be substantially higher than the "likely" range in the 21st century if the sections of the Antarctic ice sheet that have bases below sea level were to collapse" (Church et al., 2013, p. 1445). For the U.S. coastline, it is estimated that sea level might be 25% to 30% higher than the average level globally, due to regional variations in earth's gravity field (Bamber et al., 2009; Mitrovica et al., 2009; Cameron et al., 2012).

In the past, most studies of climate change impacts on natural systems have focused on the impacts of rising temperatures, intensified storms and hurricanes, changes in precipitation, and climatic variability (Parmesan & Matthews 2006; Heller & Zavaleta, 2009; Noss, 2011). Recently, increasingly more attention has been paid to the effects of SLR associated with climate change (Noss, 2011, Hoctor et al., 2014; Noss et al., 2014). It is accepted that SLR is likely to cause species extinction and ecosystem disruption in the near future (Noss, 2011).

Primarily, two factors including warming sea water causing thermal expansion and warmer air temperature causing land-based ice melt are contributing to SLR (Nicholls & Cazenave, 2010). Sea level changes could vary place to place due to the combination of contributors including shifting surface winds, the addition of melting ice and the expansion of warming ocean water (Church et al., 2013). Globally, the variations of land ice distribution can affect the earth's shape and gravitational field, which could lead to regional fluctuations in sea level. At the regional scale, regional subsidence and uplift could also influence SLR rates. At the local scale, localized processes such as sediment compaction and tectonics could cause variations of sea level (Church et al., 2013).

"Due to a combination of low lying elevations, a high water table, peninsular geography, vulnerability to tropical storms, and a large and growing human population concentrated near the coasts, Florida is extremely susceptible to SLR caused by global climate change" (Noss et al., 2014, p. i). Even at SLR of only 1.0 meter, 10% of the state's lands are likely to be inundated by sea water (Noss, 2011). With higher SLR, the percentage of inundated areas along the coast will increase dramatically (Figure 1-2).

SLR affects habitats, species and ecosystems (Geselbracht et al., 2011). Most tidal wetlands that have low freshwater and sediment supplies are likely to "drown" where the wetlands' ability to accrete vertically cannot outpace SLR (Nyman et al., 1993). Salt marshes are expected to move upslope with the rising sea water (Brinson et al., 1995). However, existing and future development is expected to be a limiting factor to the potential migration of coastal plants and this situation has already been shown (Donnelly & Bertness, 2001; Feagin et al., 2005; Desantis et al., 2007). It is estimated that the most severe losses for intertidal habitat are likely to occur at sites where the coastline is not able to move inland due to steep topography, seawalls, or intensive human land uses (Galbraith, 2002). In Florida, since many species are located only within coastal areas, without contiguous natural corridors and restoration measures, habitats, species, and entire ecosystems will likely be lost (Noss, 2011).

1.1.2 Land Use Change

In addition to SLR, land use change is another major threat to biodiversity conservation in Florida. Land use change is the physical manifestation of population growth. Florida's human population, with more than 75% of its population concentrated in the counties along the coast (Noss, 2011; CSO, 2010; Wilson & Feschetti, 2010), was approximately 19.5 million in 2013 with a projected growth rate of 17.6% over 2000 – 2100, which is a much higher rate than the projected 9.7% national growth rate (U.S. Census Bureau, 2010). Based on a moderate human population growth trend, the population of Florida is likely to reach 35.8 million in 2060 (Zwick & Carr, 2006). Florida 2060, a project funded by 1000 Friends of Florida and conducted at the University of Florida's Geoplan Center shows that the state could lose 7 million acres of additional rural lands, including 2.7 million acres of native habitats, to urbanization by 2060 if Florida keeps the current growth densities and development patterns (Zwick & Carr, 2006).

In addition to population growth, population shifts will also lead toland use change. People living in some coastal areas are already experiencing SLR impacts such as heavier flooding, higher storm surge, increased coastal erosion and saltwater intrusion (Pilkey & Yong, 2009; NRC, 2010). With higher SLR and related issues including increased storm surge, after some threshold of SLR it is likely that mass human migration from the coastal areas to inland areas will happen (Noss, 2011). This mass migration could create a second wave of native species extinction if human population migration displaces native habitat in inland areas that are not formally protected but are essential to various species of conservation interest (Hoctor et al., 2014; Noss, 2011).

Existing conservation areas in Florida will be heavily impacted by SLR and future population growth, with shifts and increasing urbanizationlikely to displace potential areas for conservation (Figure 1-3). Coupled with impacts from population growth, shifts, and increasing urban development, SLR may constitute one of the most important challenges for conserving biodiversity in Florida (Hoctor et al., 2014; Noss et al., 2014). SLR is likely to trap coastal ecosystems between "the devil and the deep blue sea", where the devil is coastal development (Harris et al., 1992; Noss, 2011). Current coastal populations are likely to migrate inland, impacting upland ecosystems, and coastal hardening (such as seawalls) will inhibit the ability of coastal ecosystems to adapt and retreat. This will by no means be a problem unique to Florida, given that nearly 44% of the global population currently lives within 150 kilometers of the coast (UN Atlas of the Oceans, 2010). The natural systems along the coast that we depend on for provision of ecosystem services, which are already under direct assault from SLR including coastal erosion and inundation, are likely to be further threatened by how we respond to SLR.

1.2 Conservation Planning and Reserve Design

Setting aside protected areas to conserve natural values of the earth has been both an ancient and widespread human activity (Margules & Pressey, 2000). Additional reserves are being added to the existing protected areas to provide more effective biodiversity conservation including species, populations, communities and ecosystems (Anon, 1992; Margules & Pressey, 2000). In Florida, approximately 9,500,000 acres of public lands are managed for conservation purposes and approximately 570,000 acres of lands are protected by conservation easements, which account for 20.8% and 1.3% respectively of the total land and water in Florida(Florida Department of Environmental Protection, 2011).

Conservation planning is the process to allocate suitable areas for promoting the persistence of natural values including biodiversity and ecosystem services (Pressey et al., 2007). One of the primary issues of conservation planning is that it is often not systematic and the identified new reserves are often not contributing to the full representation of biodiversity, protection of viable populations, or protection of sustainable ecosystems (Margules & Pressey, 2000). A primary reason is that land conservation is usually avoided in any areas where the extraction of natural resources or land development would be economically valuable (Robert, 1994; Margules & Pressey, 2000). To effectively achieve conservation goals, a new discipline called systematic conservation planning (which can also be called "reserve design") emerged in the field of conservation biology since 1980s (Pressey & Bottrill, 2008; Margules & Pressey, 2000). Systematic conservation planning can be defined as a structured, target-driven approach that aims to meet two objectives: 1) representativeness, that is the need to represent all biodiversity features; and 2) persistence, that is the need to promote long-term survival of biodiversity features (e.g., species, natural community) under all circumstances including climate change (Jantke, 2011; Margules & Pressey, 2000). Systematic conservation planning has been summarized by Margules & Pressey (2000) as a process of six stages. The starting point is to compile biodiversity data of the study area. This includes reviewexisting datasets and collect new information to augment or replace existing datasets. Secondly, conservation goals for the study area need to be identified. This stage includes setting quantitative targets for conserving species and other biodiversity features, setting quantitative targets for design criteria including minimum size and connectivity. Thirdly, a review of existing conservation areas should be conducted. This includes assessing the existing conservation areas and identifying threats to under-represented biodiversity features. In the following stage, additional conservation areas are selected. This stage includes identifying new conservation areas using reserve selection algorithms or conservation planning software to achieve conservation goals. The fifth stage is to implement the recommended conservation actions. This includes choosing the most feasible management strategies for the conservation areas and looking for alternatives if the selected areas are difficult to protect effectively. After implementation, the final stage is to maintain the ecological value of the selected conservation areas. This stage includes implementing strategies at each conservation area, monitoring the success of the strategies and modifying the strategies as needed.

Spatial conservation prioritization, a relatively new technique, has emerged in the field of systematic conservation planning in conservation biology. Spatial conservation prioritization is the use of quantitative methods to produce spatially-explicit information on conservation priorities (Moilanen et al., 2009). The effective conservation of biodiversity is constrained by limited resources, and hence the investment in conservation must be prioritized to achieve the most efficient results (Wilson et al., 2009). Spatial conservation prioritization involves Geographic Information Systems (GIS)—based decision support for guiding conservation planning (Lehtomaki & Moilanen, 2013; Ferrier & Wintle, 2009). The aim of conservation prioritization is to answer questions about when, where and how to achieve conservation goals efficiently (Lehtomaki & Moilanen, 2013; Pressey et al., 2007). Since its origin in the early 1980s, the field of spatial conservation prioritization has influenced the conservation planning process, land conservation policies, and the creation and management of new conservation areas (Moilanen et al., 2009).

Based on classic decision theory, a generic framework for setting conservation priorities is provided (Moilanen et al., 2009). In this framework, there are two broad classes of prioritization, the minimum-set and the maximum-cover approaches. The minimum-set approach is to minimize costs while achieving the targeted conservation goals. In contrast, the maximum-cover approach is to maximize the "benefits" of conservation while given a fixed budget (Moilanen et al., 2009). Widely used conservation prioritization tools such as Marxan (Ball et al., 2009), Zonation (Moilanen et al., 2009) and C-Plan (Pressey et al., 2009) are developed based on either the minimum-set approach or the maximum-cover approach.

Prioritizing assets, locations and actions are the three conservation prioritization problems that need to be solved. For prioritizing assets, the question is how much should we invest in the conservation of one species over others or in protecting one area versus others. A research project compared the use of different types of species (e.g., umbrella, keystone, flagship, indicator and focal species) in setting conservation priorities and found that using each type of species has their own strengths and weakness in priority setting (Moilanen et al., 2009). Identifying conservation priorities based on a subset of species may lead to bias because the selection of species could be unrepresentative of biodiversity conservation needs as a whole (Wilson et al., 2009).

For prioritizing locations, the question is how can we expand existing conservation areas or identify new conservation areas that have the greatest potential to achieve relevant biodiversity conservation goals. Prioritizing locations are typically based on the measure of biodiversity importance (Redford et al., 2003; Brooks et al., 2006). Many approaches such as biodiversity hotspots and key biodiversity areas have been used to prioritize locations for conserving biodiversity (Moilanen et al., 2009). The Florida Ecological Greenways Network (FEGN), which is a state plan intended to guide conservation related to protecting large, intact landscapes and wildlife corridors, is partitioned into 6 priority classes based on certain criteria (e.g., potential importance of maintaining or restoring populations of wide ranging species such as the Florida black bear and Florida panther) measuring biodiversity conservation significance (Hoctor et al., 2000). For prioritizing actions, the question is what actions (e.g., purchasing what land, managing species populations, planting trees, etc.) should be selected to best achieve conservation goals.

The Marxan and Zonation software packages have been developed for selecting conservation priorities based on spatially-explicit biodiversity data (e.g., species habitat distribution, natural communities distribution) using mathematical algorithms (Moilanen et al., 2009). The Marxan software package was developed at the University of Queensland based on the minimum-set approach to identify conservation priorities that achieve conservation targets at near minimum cost (Delavenne et al., 2012; Possingham et al., 2000). In contrast, Zonation software was developed at the University of Helsinki based on the maximum-cover approach to identify conservation priorities that maximize conservation benefits under a fixed conservation budget (Delavenne et al., 2012; Moilanen et al., 2009). Though both Marxan and Zonation have been used widely in decision support of spatial conservation prioritization, there are differences between the outputs of these two methods. Delavenne (2012) compared the outputs of Maxant and Zonation in marine protected areas (MPA) selection in the Eastern English Channel and found that both software packages identified similar conservation outcomes with Marxan generally producing more efficient conservation results while Zonation produced results that emphasized ecological connectivity.

1.3 Sea Level Rise and Spatial Conservation Prioritization

Climate change challenges conventional conservation planning and reserve design approaches because climate change is likely to alter the suitable areas for species conservation (Peters & Darling, 1985; Araujo, 2009). As sea level rises, many coastal species and habitats are likely to move upslope with rising sea water (Brinson et al., 1995), but current and future development is likely to limit this migration (Donnelly & Bertness, 2001; Feagin et al., 2005; Desantis et al., 2007). Existing conservation areas along the coasts are under direct threat of inundation from rising sea water and the potential threat includes insufficient protection of imperiled species. To facilitate biodiversity adaptation to this environmental change, projected SLR needs to be incorporated into conservation planning and the reserve design process to assess potential changes in conservation priorities and provide spatial recommendations for avoiding, minimizing, mitigating, and facilitating adaptation to SLR.

In low-lying coastal regions as found in most of Florida, identification of conservation priorities in response to SLR and land use change has become an urgent task for conservation planners and coastal decision-makers. In January 2010, a symposium held by conservation biologists at the Archbold Biological Station in Florida identified the research needs for adaptation to SLR in Florida. Noss (2011) summarized the conservation planning needs in the symposium and here are the two that are most relevant to spatial conservation planning.

- “Spatially-explicit identification of where coastal development currently presents barriers to species migration and natural community shifts and where projected development over the next few decades is likely to create new or more severe barriers" (Noss, 2011, p. 10).

- "Identification of areas where existing conservation areas should be enlarged (i.e., boundaries expanded inland), where corridors should be established from coastal to inland sites, and where new conservation areas (e.g., inland refugia and marine/estuarine reserves) should be created" (Noss, 2011, p. 10).

Though the biological side of conservation planning has advanced rapidly, there is little attention paid to the economic side (costs and benefits) of conservation planning (Naidoo & Ricketts, 2006). As funds and other resources are usually severelylimited, investments in conservation must be strategically and economically allocated to achieve efficiency (Margules & Pressey, 2000). Despite the fact that conservation planners have been slow to incorporate economic analysis into conservation practices (Naidoo et al., 2006), some studies have demonstrated the importance and necessity of applying economic concepts and tools in conservation decision-making (Watzold et al., 2006; Naidoo et al., 2006). Incorporating economic considerations into SLR adaptive conservation prioritization is useful because coastal reserve managers and other conservation decision-makers will need to know where and how to allocate their limited funds and resources efficiently to achieve conservation goals.

Conservation prioritization requires spatial data about the distribution of biodiversity features such as species, habitat and natural communities (Regan et al., 2009; Moilanen et al., 2009). To identify conservation priorities under SLR, one of the greatest challenges for planning is predictingthe distribution of future biodiversity features. The decision-making environment associated with SLR and climate change is characterized by high uncertainty (Noss, 2011). Uncertainty exists about selecting the appropriate coastal impact models and elevation datasets potentially resulting in different outcomes (McLeod et al., 2010; Runting et al., 2013). Coastal impact models vary because they have different level of complexity and requirements for data accuracy, from the simplistic "bathtub" model that assumes everything will be inundated below a certain water level, to the more complex models such as Sea Level Affecting Marshes Model (SLAMM; Clough et al., 2010) that takes ecological processes of habitat transition into account (McLeod et al., 2010).

High resolution elevation data (e.g., Lidar-based) and more complicated models (e.g., SLAMM) are preferably used in SLR adaptation planning to predict marsh habitat change and provide foundations for identifying conservation priorities in response to SLR and land use change (Runting et al., 2013). Nevertheless, we know little about the effectiveness of the improved dataset-model combination in SLR adaptive conservation designs and to what extent this improved dataset-model combination could enhance SLR adaptive conservation outcomes. This is important to know because high resolution data are likely not available in some data-poor coastal areas and running the more complicated models is time-consuming, expensive and requires expert knowledge.

To meet these research needs, this dissertation develops an integrated conservation planning process that utilizes spatial prioritization tools and incorporates economic concepts to identify conservation prioritization in the face of landscape dynamics due to SLR and land use change, and compares the conservation outcomes with different coastal impact model and elevation dataset combinations. Results and recommendation from this research could provide coastal reserve managers and other conservation decision-makers analysis options and spatially-explicit informationon how to facilitate biodiversity adaptation to SLR. The conservation prioritization process and results of this research can be used by conservation decision-makers to seek opportunities for additional conservation lands acquisition to facilitate adaptation to SLR and land use change. Objectives of this dissertation include:

- Developing an integrated conservation planning process to identify SLR adaptive conservation prioritization. Such conservation prioritization should be useful to reserve managers for assessing conservation and land use options in response to SLR and land use change.
- Incorporating economic concepts into the SLR adaptive conservation planning process to identify conservation prioritization that could achieve greater conservation efficiency. This research informs coastal decision-makers about where to allocate their limited funds and resources efficiently to achieve conservation goals.
- Investigating the value of elevation data and geomorphology model accuracy in SLR adaptive conservation prioritization. Findings from this research could assist conservation decision-makers to better understand the value of investing in more accurate datasets and models when designing adaptation plans for biodiversity conservation in low-lying coastal regions such as Florida and conservatively interpreting the results of lower resolution data and lower complexity models.

- Making policy recommendations for conserving biodiversity in the face of landscape dynamics due to SLR and land use change. This includes recommendations for incorporating SLR adaptive conservation priorities into local land use plans and changing growth patterns including increasing residential densities to avoid or minimize future development on lands with conservation value.

1.4 Study Area

Florida is an instructive case study for SLR adaptation, and how Florida responds to SLR will offer lessons to other coastal regions in the world (Noss et al., 2014; Noss, 2011). The ecological and social-economic features of the Matanzas River Basin in Florida provide a salient case study for developing adaptive conservation designs in response to SLR and land use change, so this area has been selected as the case study for this dissertation. To incorporate more regional land use and ecological considerations (e.g., the forests adjacent to the west side of the basin are within the Florida Ecological Greenways Network), the study area includes a 5 kilometers buffer beyond the Matanzas River Basin (Figure 1-4).

Located along Northeast Florida's Atlantic coastline, the Matanzas River Basin is one of the most valued and threatened areas along the Florida coastline. The Matanzas Riveris a coastal estuarine water body that extends from the St. Augustine Inlet southward into the Intracoastal Waterway south to Palm Coast, Florida. The basin covers approximately 100,000 acres between the City of St. Augustine and the City of Palm Coast and a large area of rural lands to the west. The Matanzas River basin had nearly 90% of its land in undeveloped natural or rural condition, thus providing a rare opportunity to incorporate SLR into future conservation and land-use plans with little conflict with existing development. The basin includes the Matanzas River estuary and associated coastal forests, marshes, wetlands, uplands and beaches (Matanzas River Basin Biodiversity Initiative, 2008). The basin contains the southern component of the Guana Tolomato Matanzas National Estuary Research Reserve (GTMNERR), the Fort Matanzas National Monument and other Florida conservation management areas including the Matanzas State Forest and the Faver-Dykes State Park.

The Matanzas River Basin ecosystem is home to many species of plants, animals, fungi and microorganisms distributed among various habitats in the basin. Major habitat types in the basin include upland forest (approximately 55%), saltwater marsh, freshwater marsh, and open water. In the GTMNERR, more than 40 mammal, 350 bird, 40 reptile, 20 amphibian, 300 fish and 580 plant species have been documented and many more species are expected to occur. Among these species, 48 are protected animals and 8 are protected plants with federal or state listing statusincluding gopher tortoise (Gopherus polyphemus), Florida pine snake (Pituophis melaneleucus), gopher frog (Rana capito), Anastasia Island beach mouse (Peromyscus polionotus phasma), etc. (GTMNERR management plan, 2009). Because the basin is relatively flat, SLR could have strong impacts on coastal areas that provide habitat for these species.

Existing conservation layers in Florida identified large areas of lands important for conservation within the Matanzas study area. The Florida Ecological Greenways Network (FEGN; Hoctor et al., 2013), which identified opportunities to protect large, intact landscapes that are critical for conserving biodiversity and ecosystem services in Florida, identified large areas as priority 2, 4 and 6 in the Matanzas study area, where 2 is the highest priority in the three (Figure 1-5). In addition, the Critical Lands and Waters Identification Project Dataset (CLIP; Oetting et al., 2014), which is statewide spatial data that could be used to identify conservation priorities for a broad range of natural resources, also identified large areas as conservation priorities within the Matanzas study area (Figure 1-6). Both FEGN and CLIP take SLR into account in a general way, but do not yet address specific impacts and potential changes in habitat priorities for focal species based on SLR.

Located within one of the fastest growing regions in Florida, the Matanzas River Basin is under great pressure from future development. The population of St. Johns and Flagler Counties, where the Matanzas River basin is located, has grown approximately 40% since 1990 (Florida Department of Environmental Protection, 2012). More than 1.5 million people are living within 50 miles of the basin including the cities of Jacksonville, St. Augustine, and Daytona Beach. Based on Florida 2060 report (Zwick & Carr, 2006), it is estimated that developed land uses in this region could increase significantly if current population growth rates continue in the coming decades.

Under the combined impacts of SLR and land use change, there is an urgent need to develop adaptive conservation strategies for biodiversity conservation in the Matanzas study area. To plan for SLR adaptation, a project led by Dr. Kathryn Frank in the College of Design, Construction and Planning at the University of Florida is collaborating with the GTMNERR to assess the impacts of SLR and land use change on species and natural communities and to make recommendations for conserving natural systems within the basin. The regional conservation priorities developed for the Matanzas SLR adaptation planning project were based on existing conservation data aggregation. To fully integrate SLR considerations in the planning process, the use of a reserve design algorithm to identify conservation priorities under SLR scenarios could add useful information beyond the conservation priorities identified by the project team. Methodologies developed in this research should be transferable to other low-lying coastal areas around the world for SLR adaptation planning.

CHAPTER 2
CONSERVATION PRIORITIZATION IN RESPONSE TO SEA LEVEL RISE AND LAND USE CHANGE

2.1 Coastal Habitat Change Due to Sea Level Rise

SLR is likely to constitute the most important challenge for biodiversity conservation in the Matanzas study area when coupled with interacting impacts from human population growth and increasing urbanization. To plan for SLR, a project led by the University of Florida and the GTMNERR, and funded by the National Estuarine Research Reserves (NERRS) Science Collaborative, assessed SLR impacts on natural and human communities and recommended adaptation options in response to SLR and land use change in the Matanzas River Basin.This study provided important building blocks for my research including a detailed SLAMM modeling effort that made SLR projection-based species habitat models more resolute.

For this interdisciplinary SLR adaptation planning project, Dr. Anna Linhoss and Dr. Greg Kiker in the Department of Agricultural and Biological Engineering at the University of Florida used SLAMM to simulate SLR and wetland migration in the Matanzas study area (Linhoss et al., 2014). SLAMM has been widely used to simulate long term SLR impacts on shorelines and wetlands across the nation since its development in the mid-1980s (NOAA, 2014). In Florida, Dr. Greg Kiker at the University of Florida has run SLAMM for all Florida's coastline except the Everglades and the Florida Keys (Hoctor et al., 2014). Previous work using SLAMM has described its advantages and disadvantages and has generally shown that SLAMM is very competent in its simulations (Mcleod et al.,

2010). The advantages of SLAMM include its flexibility in scale, ability to describe the vulnerability of species habitat,and ability to simulate the effect of saltwater intrusion on land cover (Mcleod et al., 2010).

The inputs for SLAMM include the elevation dataset, land cover data, site specific information (e.g., accretion rate, storm frequency, tidal trends, erosion rate, etc.), and future SLR scenarios (Aiello-Lammens et al., 2011). For this study, the model was run in a participatory setting where involvement and feedback from the academic study team as well as GTMNERR personnel was integral to determining model parameters and interpreting theresults (Linhoss et al., 2014). This participatory approach was useful in which it gave ownership of the model results to all of those involved, helped to identify additional sources of input data, and ensured that the model was being used in the manner best suited to meet all of the objectives of the team.

Based on the detailed parameters, the SLAMM was run with 0.2, 0.5, 1.0, 1.5, 2.0, and 2.5 m SLR projections in the Matanzas study area. The output maps (Figure 2-1) and tables (Table 2-1) for 0.5, 1.0 and 2.5 m SLR show that the developed dry land is especially vulnerable to SLR and up to 8% of these areas would be lost toa 1.0 m SLR and up to 25% of these areas would be lost to a 2.5 m SLR. Additionally, the inland freshwater marsh is also shown to be vulnerable to SLR and is likely to lose a total of 7% of their area under the 2.5 m scenario. However, transitional saltmarsh is expected to increase 81.4% under the 1.0 m scenario but only 15.6% under the 2.5 m scenario. As sea level rises, beaches are shown to increase in area and could increase up to 137% of their area under the 2.5 m scenario. However, for this to occur, existing developed land will be converted to beach. Because of the value of beachfront property, it is unlikely that these beaches will be allowed to migrate onto developed lands. On the other hand, beach nourishment is a common practice in Florida and nourishment activities will likely keep the beach areas near their current locations at least for the near future (Linhoss et al., 2014).

Species habitat will change along with land cover change due to SLR. Estuarine species such as the American oystercatcher and Black rail are likely to change significantly because of the transition between wetlands and open water after SLR in the Matanzas study area. To predict future habitat for species under SLR, species habitat models are needed. In recent years, species distribution modeling has been used widely to estimate the actual or potential distribution of species and a variety of modeling techniques have been developed for this purpose (Pearson, 2007; Guisan & Thuiller, 2005). Two types of input data are needed for modeling species distributions including the species occurrence records or range information and relevant environmental variablessuch as land cover, soil types, elevation, etc. (Pearson, 2007).

Species habitat is one of the biodiversity features that could be used in spatial conservation prioritization tools to identify areas of high ecological value (Moilanen et al., 2009). Future projected species habitat under various SLR scenarios in the Matanzas study area could serve as appropriate planning inputs to identify conservation prioritization in response to SLR and land use change. Using potential species habitat under SLR as input, this chapter uses a spatial conservation prioritization tool to identify conservation priorities in the face of landscape dynamics due to SLR and land use change. The SLR adaptive conservation priorities identified in this research will inform coastal decision-makers "where existing conservation areas should be enlarged, where corridors should be established from coastal to inland sites, and where new conservation areas (e.g., inland refugia and marine/ estuarine reserves) should be created" to facilitate biodiversity adaptation to SLR (Noss, 2011).

2.2 Methods

A spatially explicit modeling process that incorporates geomorphological modeling (SLAMM), species habitat modeling (Species Habitat Models) and conservation prioritization modeling (Zonation software) was used to identify a system of areas that will protect biodiversity in response to future SLR and land use change in the Matanzas study area (Figure 2-2).

2.2.1 Sea Level Rise Scenarios

SLR scenarios are fundamental to vulnerability assessments and all other parts of the adaptation planning process in coastal areas. For this research, I chose three potential SLR scenarios for 2100 (0.5 m, 1.0 m, and 2.5 m)to identify adaptive conservation designs. The 0.5 m SLR projection is the lowest case scenario and it falls in the "likely" SLR range projected by AR5. The 1.0 m SLR projection is the intermediate scenario and it islikely the worst case SLR projection in AR5. I chose 2.5 m as the highest SLR scenario by 2100 because this was the highest SLR projection used in SLAMM modeling for the Matanzas SLR adaptation project led by Dr. Kathryn Frank and this extreme case is possible "if the Antarctic ice sheet were to collapse" (Church et al., 2013, p. 1445).

2.2.2 Predicting Species Habitat Using Species Habitat Models

In Florida, Dr. Tom Hoctor from the Center for Landscape Conservation Planning at the University of Florida, the Florida Fish and Wildlife Conservation Commission (FWC) and the Florida Natural Areas Inventory (FNAI) have developed most of the species habitat models for focal species (See Appendix A, which shows the modeling scripts for American oystercatcher as an example).These models have been used to conduct an assessment of vulnerability of Species of Greatest Conservation Need (SGCN) to SLR in Florida (Hoctor et al., 2014; Noss et al., 2014). Using species habitat models for assessing species vulnerability to SLR has several advantages. Firstly, one of the primary impacts of SLR on coastalnatural systems is habitat change especially shoreline and wetland change, and we can predict this change using SLAMM. Species habitat models which predict potential species distributions based on their ecological affinities, land cover, and other relevant spatial data are good at reflecting this change. Secondly, such species habitat models do not require species occurrence data as inputs. We have species occurrence data for some species at the current condition (before SLR) but we do not have species occurrence data for the future condition (after SLR). Thirdly, the species habitat models from Hoctor were developed specifically for local focal species in Florida complying with the Florida Land Cover and Use Classification System (FLCUCS) and thus could have better accuracy compared to other generally used models. Finally, compared to other models, Hoctor, FWC and FNAI species habitat models are relatively easy to use. All of these models also were developed in collaboration with relevant species experts and have been peer reviewed (Hoctor, 2011).

The inputs for running species habitat models depend on each species' specific habitat requirements. Land use data is the primary input and other data inputs may include soil types, hydrology data (major rivers, streams) and nesting sites. Revised FLCUCS land use data from the St. Johns River Water Management District (SJRWMD) modified with more detailed natural community data from both the Florida Cooperative Land Cover data and GTMNERR coastal wetlands data were used to identify current species habitats based on their land cover affinities and other habitat requirements such as suitable soils. A revised land use dataset incorporating SLAMM results was used to rerun these models to identify future habitats under different SLR scenarios. For the Matanzas study area, 38 focal species including bald eagle, gopher tortoise and sea turtle were selected by Dr. Tom Hoctor for species habitat modeling under the 0.5 m, 1.0 m and 2.5 m SLR scenarios (Appendix B).

The habitat scripts were run within ArcMap 9.3 of Environmental Systems Research Institute (ESRI) using raster calculator. The resulting species habitat grids had a cell size of 10 m. The cell values of each species habitat could vary from 0 to 8 with 0 representing unsuitable habitat and all other positive values representing suitable habitats for that species in either nominal or prioritized classes depending on the species.

2.2.3 Landscape Prioritization with Zonation

The Conservation planning software Zonation (Moilanen & Kujala, 2006) was developed for spatial conservation prioritization based on observed or predicted distribution of biodiversity features (e.g., species, habitat types). This software has been used to identify conservation prioritiesin response to SLR in a coastal region of South East Queensland, Australia (Runting et al., 2013). Zonation can be used to identify areas important for retaining both habitat quality and connectivity for multiple species or other biodiversity features, thus providing conservation and land use decision makers a quantitative method to protect biodiversity in the long run (Molianen et al., 2012). Major outputs of Zonation include: 1) a ranking of conservation priority throughout the selected study area and 2) curves (x-y plots) that describe the performance of species at different levels of landscape removal. Typically, Zonation can be used to 1) assess existing and proposed conservation areas; 2) expand existing conservation areas; 3) identify new conservation areas to achieve certain conservation goals.

The Zonation algorithm is characterized as a reverse stepwise heuristic algorithm because it "iteratively removes cells whose loss will cause the smallest marginal loss in the overall conservation value of the remaining landscapes while maintaining connectivity" (Molanen et al., 2009, p. 197). The principle on which the Zonation algorithm is based is that minimizing the loss of conservation value will result in the greatest conservation value in the remaining areas (Gordon et al., 2009). There are different rules determining the marginal loss of conservation value used in the Zonation algorithm (Molianen et al., 2009). Moilanen and others (2012) summarized that presently there are four conceptually different cell removal rules: 1) core-area Zonation (Moilanen et al., 2005), which gives the highest values to the most important locations within each species distribution;2) concave additive benefit function (Moilanen, 2007), which treats occurrences within a species as additive, that is, sites containing more species tend to receive higher values than species – poor sites;3) target-based benefit function (Moilanen, 2007), which is implemented using a modified benefit function, and 4) the generalized benefit function (Moilanen et al., 2012), under which cells with the lowest marginal value summed across all species will be removed. There are significant differences among these cell removal rules and the preferable rule to be chosen depends on the goals of the planning work (Moilanen et al., 2012). I have chosen "core-area Zonation" for identifying landscape prioritization in the

Matanzas study area as this rule is most appropriate when there is a set of species all of which are to be protected and tradeoffs between species are discouraged (Moilanen et al., 2005; Moilanen et al., 2012). For the selected 38 focal species in the study area, the goal is to identify conservation prioritization obtaining high-quality areas for all of these species thus "core-area Zonation" is an appropriate rule for this analysis.

Besides selecting high-quality areas for conservation, Zonation emphasizes connectivity in the reserve selection process. Moilanen (2012) recommended using aggregation methods in reserve design as the benefits of increased connectivity is usually high compared to the cost of losing a minor amount of biologically valuable areas. Zonation has some distinctive aggregation methods, which include boundary length penalty (Moilanen & Wintle, 2007), distribution smoothing (Moilanen, 2005; Moilanen& Wintle, 2006), boundary quality penalty (Moilanen & Wintle, 2007), directly connectivity (Moilanen et al., 2008) and matrix connectivity (Lehtomaki et al., 2009), to achieve the goal of increasing connectivity. Choosing the right aggregation method depends on the specific conservation targets and computational issues, and it is recommended not to use multiple aggregation methods simultaneously in order to avoid difficulties for interpreting results (Moilanen et al., 2012).

To increase connectivity in the Matanzas study area, species-specific connectivity is integrated into Zonation through the distribution smoothing method built into the Zonation software. Distribution smoothing is a species-specific aggregation method that retains well connected areas, thus leading to a more compact solution (Moilanen et al., 2012; Moilanen & Wintle, 2006; Moilanen, 2005). According to the Zonation user manual (Moilanen et al., 2012), this aggregation method does a two-dimensional calculation that identifies areas of high habitat quality and density. Using the distribution smoothing method, we can effectively identify important semi-continuous areas where the species has overall high levels of occurrence (potential habitat), although not necessarily in every grid cell (Moilanen et al., 2012).

To include distribution smoothing in the analysis, the width of the dispersal kernel for each species need to be determined. The degree of smoothing for each species α is estimated by the dispersal capability of the selected species using this calculation:

$$\alpha=2 \text{ / Use of landscape (in same units as cell size)}$$

The "use of landscape" is the average dispersal capability of species and the units of this number should be consistent with the cell size units of the biodiversity feature grids.

Inputs for running distribution smoothing in Zonation include:

- A set of biodiversity feature grid layers. For the Matanzas research, these are the grids of species habitats derived from the species habitat models.
- A biodiversity feature list file that contains a list of species or other biodiversity features (e.g., habitat types, natural communities, etc.) with set-up parameters such as species weight and α value. In this research, the biodiversity feature list is a list of the selected focal species. The weight for each species is based on FNAI state rank for individual species and the rank for the guilds is based on the rarest species in that guild. The α value for each species is calculated based on the average species dispersal capability based on expert opinion (Appendix B).
- A run settings file that contains all basic Zonation settings. Appendix C shows the run settings file used for the dissertation research.

2.3 Results

2.3.1 Species Habitat Change

Appendix D presents the change in species habitats based on SLR, projected development and a combination of SLR and land use change for the selected 38 focal species in the Matanzas study area. Based on the statistics, all species are likely to lose some habitat to projected development but not all species are likely to lose habitats due to SLR. Under all three SLR scenarios, most of the species are likely to experience habitat loss due to SLR, but some marsh dependent species could gain habitat due to SLR. Under the 1.0 m SLR scenario, black rail, Florida mink, sea turtle and beach mouse are likely to lose habitats in percentages of 58%, 45.3%, 64.2% and 43.3% respectively. However, migratory (wintering) waterfowl, American oystercatcher, the shorebird guild-open water foraging, and diamondback terrapin are likely to gain habitats in percentages of 49.7%, 167.7%, 25.5% and 35.7% respectively. Under the 2.5 m sea level rise scenario, these four species are likely to gain habitats of 159.3%, 122.7%, 78.8% and 40.4% respectively compared to their current habitats. Most species are likely to experience more habitat loss as sea level rises from 0.5 m to 2.5 m, but some species are likely to experience less impacts as sea level rises up to 2.5 m. These results indicate potentially critical thresholds in habitat changes where potential additions of more wetlandsat higher levels of SLR might offset habitat losses associated with lower levels of SLR.

Figure 2-3 shows the species habitat richness under current conditions and the 0.5 m, 1.0 m and 2.5 m SLR scenarios in the Matanzas study area. The figure provides a general overview of species habitat distribution in the study area. The species habitat richness is calculated by reclassifying each species habitat grid into 0 (not suitable habitats) and 1 (suitable habitats) and adding them in ArcGIS 9.3 raster calculator. For the 38 focal species, species habitat richness varies from 0 to 26 under the current conditions (which indicate that the maximum amount of species habitat overlap is 26). As sea level rises, species habitat richness changes in the study area.Species habitat richness varies from 0 to 21 in the landscape under the 0.5 m SLR scenario and varies from 0 to 22 under the 1.0 m and 2.5 m SLR scenarios. The highest species habitat richness areas occur around the southern portion of the GTMNERR under the three SLR scenarios, which good indicators that existing conservation lands could protect a large portion of species habitat richness even as sea level rises. The high species habitat richness areas are important for conservation as these areas are habitats for the most focal species. However, if we select conservation priorities based solely on species habitat richness, we are at risk of neglecting species whose habitat is not included in the high species habitat richness areas (Gordon et al., 2009).

2.3.2 Conservation Prioritization

Figure 2-4 shows the output produced by Zonation under current conditions and each SLR scenario with a color gradation representing the conservation value of each cell in the landscape. The red color represents sites that are most valuable for conservation and the blue color represents sites that are least valuable for conservation. Comparison of the four maps indicates that conservation priorities will change as sea level rises from 0.5 m to 1.0 m and conservation priorities will change dramatically as sea level rises up to 2.5 m. High priority areas for conservation are identified in the west-central and southeastern parts of the study area in the 2.5 m SLR scenario which are not identified in the current, 0.5 m and 1.0 m SLR scenarios.

Figure 2-5 shows the top 20% priority areas produced by Zonation with species habitat richness maps under the current conditions and each SLR scenario. The comparison indicates that a large portion of the species' habitat richness areas are also located within the high priority areas produced by Zonation. However, not all high species habitat richness areas are designated as high priority areas and some lower species habitat richness areas are designated as high priorities for conservation. The reason for this is that the Zonation algorithm selects conservation priorities based on the cell's aggregated value for conserving each species not the cell's value of species habitat richness. The ranking of each cell in Zonation is affected by the species' relative weighting and other parameters in the species list file (Appendix B).

2.3.3 Assessing the Current Reserve System

The Florida Natural Areas Inventory (FNAI) has maintained a dataset that identifies conservation lands in Florida. Figure 2-6 demonstrates the Zonation outputs under the current conditions and each SLR scenario with current designated Florida managed areas in the Matanzas study area. For all the four scenarios, a large proportion of the existing conservation lands overlap with the top 25% high priority landscapes. However, not all high priority areas identified by Zonation are included in the current conservation areas. Some areas identified as high conservation priorities for adaptation to SLR are not currently in managed areas and this situation is exacerbated when sea level rises from 0.5 m to 2.5 m. The high priority areas contiguous to current managed areas are ideal sites for expanding current reserves and the high priority areas away from the current managed areas are potentially ideal sites for creating new reserves.

Table 2-2 to Table 2-5 present the statistics of conservation prioritization identified by Zonation under the current condition and each SLR scenario with existing conservation areas. Each category of conservation priority is calculated by its area and percentage within/ not within existing conservation areas. Under the current conditions, approximately 30% of the top 2% conservation priorities identified by Zonation are not included in the existing conservation areas. More than 70% of the top 2% and more than 50% of the top 2%-5% conservation priorities under the current, 0.5 m SLR and 1.0 m SLR scenarios are within existing conservation areas. However, when sea level rises to 2.5 m, only 65% of the top 2% and 41% of the top 2%-5% conservation priorities are within existing conservation areas.

2.3.4 Assessing the updated Florida Ecological Greenways Network (FEGN)

"The Florida Ecological Greenways Network (FEGN) identifies opportunities to protect large, intact landscapes important for conserving Florida's biodiversity and ecosystem services, and serves as one of the conservation priority foundations for biodiversity and ecosystem protection efforts in Florida" (Hoctor et al., 2013, p. i). The original FEGN was delineated in 1997 and an updated FEGN was completed in 2013 with new GIS data layers identifying areas of conservation significance and representing the latest land use change. The updated FEGN has 6 priority levels for the state and has three priority levels in the study area including priorities 2, 4 and 6 with priority 2 is the highest priority in the study area (Hoctor et al., 2013).

Figure 2-7 demonstrates Zonation outputs with the updated FEGN overlaid as cross hatching. The comparison shows that a large proportion of lands with high conservation value identified by Zonation are also included in the updated FEGN priorities. Under the current conditions, there are approximately 74% of the top 2% and 85% of the top 5% conservation priorities identified by Zonation are within FEGN priorities. Under the 0.5 m, 1.0 m and 2.5 m SLR scenarios, there are approximately 72%, 71% and 70% of the top 2% conservation priorities identified by Zonation are within FEGN priorities respectively. However, there are still approximately 30% of the top 2% and 12% of the top 2%-5% conservation priorities not included in the updated FEGN under the current condition and each SLR scenario. Table 2-6 to Table 2-9 presents the statistics of the overlay between each SLR adaptive landscape prioritization and the updated FEGN. The results indicate that the updated FEGN could be used as a foundation to identify adaptive strategies in response to SLR. However, additional high priority lands will need to be identified for biodiversity conservation in response to SLR in Florida,since the FEGN does not include all coastal areas that might be important for regional to local conservation adaptation.

2.3.5 Assessing the Critical Lands and Waters Identification Project (CLIP) Database

"The Critical Lands and Waters Identification Project (CLIP) is a collection of spatial data that identify statewide priorities for a broad range of natural resources in Florida" (Oetting et al., 2014, p. 2).Through a collaborative effort between the Florida Natural Areas Inventory (FNAI), the University of Florida Center for Landscape Conservation Planning and the Florida Fish & Wildlife Conservation Commission (FWC), CLIP is organized into a hierarchical database that is used to inform long range planning decisions (Oetting et al., 2014). CLIP was initiated in 2006 by the Century Commission for a Sustainable Florida and the latest version CLIP 3.0 was released in February 2014. CLIP 3.0 has an aggregated set of priorities with five priority levels that depicts conservation significance for protecting biodiversity, landscape attributes, and high quality surface water at the statewide scale with priority 1 representing the highest priority for any of the three resource categories (Oetting et al., 2014). Figure 2-8 shows the Zonation outputs with CLIP 3.0 as simple hatching in the Matanzas study area. The comparison demonstrates that a large proportion of CLIP 3.0 priorities are also priority areas produced by Zonation. Table 2-9 to Table 2-12 present the statistics of the overlay between Zonation outputs and CLIP top 3 priorities. The results show that more than 90% of the top 10% conservation priorities identified by Zonation are included in the CLIP top 3 priorities under the current condition and all SLR scenarios. This reveals that the CLIP database could be used as a very good foundation to identify conservation priorities for adaptation to SLR in Florida.

2.3.6 Assessing Projected Development

Florida 2060, a project funded by 1000 Friends of Florida and conducted at the University of Florida's Geoplan Center shows that the state could lose 7 million acres of additional lands including 2.7 million acres of native habitats to urbanization by 2060 if Florida keeps the current growth and development pattern (Zwick & Carr, 2006). Specifically for the Matanzas study area, Dr. Paul Zwick in the Department of Urban and Regional Planning at the University of Florida has projected future development by 2060 based on a combination of 1.0 m SLR, current population shift from areas directly threatened by a 1.0 m SLR, and future projected human population growth by 2060. Figure 2-9 shows the Zonation outputs with development projection developed specifically for the Matanzas study area in 2060 as cross hatching under each SLR scenario in the study area. Table 2-14 to Table 2-16 present the statistics of the overlay between each SLR adaptive landscape prioritization and the projected development. The results demonstrate that there will be 6.2%, 6.0%, 5.4% of the top 10% conservation priorities projected to be converted into development under the 0.5 m, 1.0 m and 2.5 m SLR respectively. This reveals that if no action is taken, some conservation priorities that could be acquired for adaptation to SLR may be lost to future development.

2.4 Discussion

Figure 2-4 shows conservation priorities identified by Zonation under current conditions and the three SLR scenarios in the Matanzas study area. Some areas were identified as top conservation priorities in all four scenarios and some areas were only priorities under one or two scenarios. Conservation priorities identified by Zonationshift as sea level rises. This is because habitat for each species —the primary input data for running Zonation, will change as land cover changes based on the SLAMM results. Zonation identified a large new area of high conservation priority in the western part of the study area most likely because inland habitat for some focal species was needed to replace lost habitat for those species in coastal areas directly affected by SLR. The Figure 2-10 demonstrates the change of top 10% conservation priorities for each of the four scenarios and shows where the priorities overlap and where they do not.

The change of conservation priorities provides critical information for coastal reserve managers to develop conservation strategies for adaptation to SLR. The overlap areas of high priorities under the four scenarios indicate that these areas are "no regrets" areas for conservation. They are not only highly valuable to conserve biodiversity under the current conditions but also highly valuable to facilitate biodiversity adaptation to future SLR thus these areas need to be protected as high priority across scenarios. Based on the map of conservation priority change, coastal decision-makers will know the different strategies that they need to take to conserve biodiversity in response to different SLR scenarios.

Existing managed areas contain a large number of critical habitats for focal species in the study area. This is indicated by Figure 2-3 that most areas with high species habitat richness in our study area are within managed lands. Figure 2-6 demonstrates that there are large overlaps between Zonation conservation priorities and current managed areas. However, not all high priority areas identified by Zonation are included in the existing conservation areas. To facilitate biodiversity adaptation to SLR, additional conservation lands will need to be acquired. Figure 2-11 presents the top 10% priority areas outside of existing conservation areas under each SLR scenario, which could be acquired for enlarging existing reserves and creating new reserves based on Zonation results. To protect the top 10% conservation priorities, approximately 11,700, 10,900 and 15,200 acres of land need to be acquired for the 0.5 m, 1.0 m, and 2.5 m SLR scenarios respectively. More lands are needed to be purchased for the 2.5 m SLR scenario because new high conservation priority areas areidentified in the western and southeastern part of the study area, which are not identified in the other SLR scenarios based on Zonation results. Figure 2-12 presents the potential areas that need to be protected around the GTMNERR under the 0.5 m, 1.0 m and 2.5 m SLR scenarios. These areas are selected based on the top 10% conservation priorities around the GTMNERR that produced by Zonation. When GTMNERR seeks opportunities to enlarge current reserve to facilitate biodiversity adaptation to SLR, these areas are potentially most significantto be acquired for conservation.

The latest versions of FEGN and CLIP have large overlaps with conservation priorities identified by Zonation. Figure 2-7 demonstrates the overlaps between Zonation conservation priorities and the updated FEGN within the Matanzas study area. The overlap may be because the updated FEGN included SLR as an important consideration when identifying the largest intact and functionally connected lands containing features of conservation significance across the state (Hoctor et al., 2013). Figure 2-8 presents the overlaps between Zonation conservation priorities and the latest version of CLIP. The overlap is large because the CLIP dataset is a broad synthesis of natural resources GIS data which can be used to support comprehensive identification of conservation efforts (Oetting et al., 2014). To augment state conservation priority layers such as FEGN and CLIP, one option for a more comprehensive adaptation approach that emphasizes ecological connectivity would be to integrate the specific priorities identified in this work into the FEGN and CLIP database since they align well and capture conservation priorities at the landscape to regional landscape scale when combined.

Figure 2-9 shows that there will be conflict between SLR adaptive conservation priorities and future development scenarios. This is because some species habitat near existing cities or major roads also is highly suitable for future urban development, though the future impact of additional development could be avoided or minimized with efforts to change growth patterns including increasing residential densities. One practical challenge for conservation planners is to develop ecologically meaningful methods that can be applied to conservation decision making and operational land use planning (Moilanen et al., 2011). With population growth and economic development, alternative land uses such as commercial and residential development are likely to compete with lands that are important for conservation. Finding a solution where land use is balanced between conservation and other land uses is a practical issue that needs to be addressed in long range planning (Moilanen et al., 2012). Computationally, this can be achieved by using a mask file that masks out future development scenarios from the landscape at the start of the Zonation analysis process (Moilanen et al., 2012). Figure 2-13 presents the outputs of Zonation that uses the future development scenario as a mask layer in the analysis while keeping other input data and parameters the same in Zonation. In this case, future conservation lands will not compete with future development as cells representing potential future development are masked out at the start of the analysis process.

Incorporating future development scenarios in the analysis will avoid conflict between conservation and development but essentially gives future projected development higher priority and does not necessarily protect some important conservation priorities in the landscape. Planning future development in a careful and smart growth oriented manner to avoid competing with conservation priorities is a key component for land use decision makers to consider. Practically, the situation showed in Figure 2-13 is more likely to happen because development usually does take priority in local land use decision making, especially in coastal areas of Florida. However, adopting proactive protection measures including changing development patterns and densities could result in protection of the top conservation priorities as shown in Figure 2-11 to facilitate biodiversity adaptation in response to SLR and land use change.

2.5 Conclusions

The state of Florida has been a leader in conservation planning and reserve design for effective protection of biodiversity (Hoctor, 2003). The state has very good databases on conservation priorities, with relevance for identifying SLR adaptation strategies including a statewide system of wildlife corridor and ecological connectivity priorities (FEGN) and other biodiversity and ecosystem services priority areas in the Critical Lands and Waters Identification Project (CLIP) database. Work to assess and revise those priorities based on potential SLR impacts has only just started and no effort has been made to explore the utility of integrating multiple conservation planning or reserve design tools with SLR impacts included.

This study is the first attempt in the state to identify SLR adaptive conservation prioritization using spatial conservation planning tools. The latest version of FEGN conservation priorities were developed by working with the Technical Advisory Group (TAG) to identify priorities based on various relevant GIS data layers (Hoctor et al., 2014). The updated FEGN has considered SLR in the identification of conservation priorities by incorporating a landscape-based SLR coastal to inland connectivity analysis (Hoctor et al., 2013). The CLIP 3.0 priorities were developed based on a collection of existing conservation data that identify statewide priorities for a broad range of natural resources including biodiversity, landscape, surface water resources, ground water resources and marine resources in Florida (Oetting et al., 2014). Different from the identification of current FEGN and CLIP priorities, Zonation identifies SLR adaptive conservation priorities based on predicted future biodiversity features (e.g., species habitat), which takes SLR into fully consideration.

Comparisons of conservation prioritization produced by Zonation with existing conservation areas and datasets in Florida suggest that existing conservation areas and existing conservation databases could protect a large proportion of SLR adaptive conservation priorities in Florida. However, there are still some top priority areas excluded in the existing conservation areas and databases as sea level rises. Comparison of conservation prioritization produced by Zonation with the future development projection shows that some top conservation priorities adaptive to SLR will be threatened by future development. In response to the combined impacts of SLR and land use change, future development projections can be integrated into the planning process to achieve the goal of balancing future development with conservation and other priorities.

Conservation prioritization tools such as Zonation can be used to assess the efficacy of existing conservation lands and to identify additional lands that need to be protected for biodiversity conservation (Gordon et al., 2009). As sea level rises, coastal reserve managers will need good information about where to expand existing reserves and create new reserves to facilitate biodiversity adaptation to SLR. In addition, urban sprawl continues to cause habitat loss, which could significantly exacerbate SLR impacts and increase the threat of extirpation or extinction for many native species (Bekessy &Gordon, 2007). This chapter presents an integrated modeling process that incorporates a geomorphological model (SLAMM), habitat models (species habitat models) and spatial prioritization software (Zonation) to identify conservation prioritization in the face of landscape dynamics due to SLR and land use change. Outputs of this analysis can be used by reserve managers and other conservation decision makers to seek opportunities for additional conservation lands acquisition.

This research provides an innovative approach for SLR adaptation planning in Florida. However, methodologies used in this research can also be transferred to other low-lying coastal areas outside of Florida to assist in conservation plans for SLR adaptation. To further advance climate change adaptation planning, especially SLR adaptation planning, additional information and research will be necessary. Firstly, the future land use datasets were created based on a conservativeassumption that existing development would stay developed as sea level rises. However, existing development might be allowed to transition to habitats as sea level rises in the future. To include this scenario, future research could identify conservation priorities based on the assumption that existing development could be converted to natural habitats as sea level rises. Secondly, aquatic species were not included in the focal species list due to model limitations. Future research could include aquatic species in the list when adequate species habitat data or models are available. Thirdly, incorporating information about the effect of SLR or development impacts on actual species or population viability (though data and time intensive) would be useful. More detailed models of ecosystem structure and function changes from SLR and climate change would be useful (such as potential changes in upland habitats from dry to more mesic, which is not addressed by SLAMM, or conversion of saltmarsh to mangrove due to temperature changes), and information on potential changes in ecosystem services as a result of climate change impacts would also be useful to advance SLR adaptation planning. Fourthly, comparing SLR adaptive conservation priorities with Zonation and other reserve design tools such as Marxan would be useful as we need to know to what extent different reserve design tools could lead to different conservation outcomes.

Table 2-1 Forecasted change in land cover categories

Land Cover Type	0.5 m	1.0 m	2.5 m
Developed Dry Land	–2.5%	–7.7%	–24.9%
Undeveloped Dry Land	–2.2%	–3.5%	–8.7%
Swamp	–0.1%	–2.5%	–7.9%
Cypress Swamp	–0.3%	–2.3%	–11.2%
Inland-Fresh Marsh	–0.1%	–2.2%	–6.9%
Transitional Saltmarsh	36.0%	81.4%	15.6%
Regularly-Flooded Marsh	–16.0%	–31.0%	13.6%
Tidal Flat	82.5%	49.0%	139.1%
Vegetated Tidal Flat	–0.1%	–14.7%	–100%
Beach	9.0%	35.4%	136.6%
Open Water	8.3%	56.9%	166.2%

Table 2-2 Statistics of conservation prioritization compared to existing conservation areas under current conditions. Percentages indicate portion of conservation priorities identified by Zonation falling within existing conservation areas

Zonation Conservation Priorities for Current Condition	Within Existing Conservation Areas (Acres)	Not within Existing Conservation Area (Acres)	Percentage of Zonation Conservation Priority within Existing Conservation
2%	3,734	1,548	70.69%
2%-5%	4,709	3,214	59.43%
5%-10%	6,688	6,517	50.65%
10%-25%	13,748	25,868	34.70%
25%-50%	8,436	57,591	12.78%
50%-75%	6,739	59,288	10.21%
75%-100%	1,517	64,461	2.30%

Table 2-3 Statistics of conservation prioritization compared to existing conservation areas under 0.5 m SLR. Percentages indicate portion of conservation priorities identified by Zonation falling within existing conservation areas

Zonation Conservation Priorities for 0.5m SLR	Within Existing Conservation Areas (Acres)	Not within Existing Conservation Area (Acres)	Percentage of Zonation Conservation Priority within Existing Conservation
2%	3,738	1,543	70.78%
2%-5%	4,120	3,802	52.01%
5%-10%	6,849	6,354	51.87%
10%-25%	13,824	25,786	34.90%
25%-50%	9,027	56,988	13.67%
50%-75%	5,841	60,174	8.85%
75%-100%	2,175	63,840	3.30%

Table 2-4 Statistics of conservation prioritization compared to current conservation areas under 1.0 m SLR. Percentages indicate portion of conservation priorities identified by Zonation falling within existing conservation areas

Zonation Conservation Priorities for 1.0 m SLR	Within Existing Conservation Areas (Acres)	Not within Existing Conservation Area (Acres)	Percentage of Zonation Conservation Priority within Existing Conservation
2%	3,964	1,317	75.06%
2%-5%	4,704	3,218	59.38%
5%-10%	6,870	6,333	52.03%
10%-25%	12,583	27,026	31.77%
25%-50%	9,361	56,654	14.18%
50%-75%	6,792	59,223	10.29%
75%-100%	1,299	64,715	1.97%

Table 2-5 Statistics of conservation prioritization compared to current conservation areas under 2.5 m SLR. Percentages indicate portion of conservation priorities identified by Zonation falling within existing conservation areas

Zonation Conservation Priorities for 2.5 m SLR	Within Existing Conservation Areas (Acres)	Not within Existing Conservation Area (Acres)	Percentage of Zonation Conservation Priority within Existing Conservation
2%	3,458	1,824	65.47%
2%-5%	3,253	4,669	41.06%
5%-10%	4,506	8,697	34.13%
10%-25%	12,478	27,131	31.50%
25%-50%	13,559	52,456	20.54%
50%-75%	7,649	58,366	11.59%
75%-100%	671	65,344	1.02%

Table 2-6 Statistics of conservation prioritization compared with the updated FEGN under current conditions. Percentages indicate portion of conservation priorities identified by Zonation falling within FEGN priorities

Zonation Conservation Priorities	Not Priorities in FEGN	Priority 2	Priority 4	Priority 6
2%	26.14%	56.23%	0	17.63%
2%-5%	14.83%	62.63%	3.22%	19.33%
5%-10%	24.10%	56.70%	4.60%	14.60%
10%-25%	22.99%	41.14%	15.60%	20.27%
25%-50%	19.42%	30.54%	33.84%	16.21%
50%-75%	20.18%	21.53%	48.71%	9.58%
75%-100%	63.46%	8.93%	26.14%	1.48%

Table 2-7 Statistics of conservation prioritization compared with the updated FEGN under current 0.5 m SLR. Percentages indicate portion of conservation priorities identified by Zonation falling within FEGN priorities

Zonation Conservation Priorities	Not Priorities in FEGN	Priority 2	Priority 4	Priority 6
2%	28.62%	56.16%	4.57%	10.66%
2%-5%	12.13%	56.82%	9.51%	21.53%
5%-10%	15.29%	58.07%	8.32%	18.32%
10%-25%	23.96%	38.94%	17.60%	19.49%
25%-50%	19.20%	31.72%	32.60%	16.48%
50%-75%	21.87%	23.31%	45.49%	9.34%
75%-100%	63.29%	7.72%	27.53%	1.46%

Table 2-8 Statistics of conservation prioritization compared with the updated FEGN under current 1.0 m SLR. Percentages indicate portion of conservation priorities identified by Zonation falling within FEGN priorities

Zonation Conservation Priorities	Not Priorities in FEGN	Priority 2	Priority 4	Priority 6
2%	29.66%	28.98%	8.40%	32.96%
2%-5%	14.01%	30.88%	12.96%	42.15%
5%-10%	23.60%	34.58%	16.19%	25.63%
10%-25%	22.82%	31.59%	28.14%	17.45%
25%-50%	23.87%	25.64%	37.21%	13.28%
50%-75%	16.91%	36.25%	39.90%	6.94%
75%-100%	62.29%	15.26%	19.90%	2.55%

Table 2-9 Statistics of conservation prioritization compared with the updated FEGN under current 2.5 m SLR. Percentages indicate portion of conservation priorities identified by Zonation falling within FEGN priorities

Zonation Conservation Priorities	Not Priorities in FEGN	Priority 2	Priority 4	Priority 6
2%	28.65%	58.50%	0	12.85%
2%-5%	12.30%	64.43%	6.33%	16.94%
5%-10%	24.97%	58.03%	6.22%	10.79%
10%-25%	21.56%	35.73%	19.84%	22.87%
25%-50%	18.28%	31.41%	33.51%	16.81%
50%-75%	22.46%	23.88%	45.72%	7.94%
75%-100%	63.09%	8.30%	26.23%	2.38%

Table 2-10 Statistics of conservation prioritization compared with CLIP 3.0 top 3 priorities under current conditions. Percentages indicate what percent of conservation priorities identified by Zonation falls within CLIP

Zonation Conservation Priorities	Within CLIP (Acres)	Not Within CLIP (Acres)	Percentage of Zonation Priorities in CLIP
2%	5,253	29	99.4%
2%-5%	7,470	453	94.3%
5%-10%	12,128	1.077	91.8%
10%-25%	35,283	4,333	89.1%
25%-50%	58,057	7,970	87.9%
50%-75%	58,201	7,826	88.1%
75%-100%	26,665	39,313	40.4%

Table 2-11 Statistics of conservation prioritization compared with CLIP 3.0 top 3 priorities under 0.5 m SLR. Percentages indicate what percent of conservation priorities identified by Zonation falls within CLIP

Zonation Conservation Priorities	Within CLIP (Acres)	Not Within CLIP (Acres)	Percentage of Zonation Priorities in CLIP
2%	5,212	70	98.7%
2%-5%	7,365	558	93.0%
5%-10%	12,466	739	94.4%
10%-25%	35,877	3,739	90.6%
25%-50%	58,174	7,853	88.1%
50%-75%	56,847	9,180	86.1%
75%-100%	27,115	38,863	41.1%

Table 2-12 Statistics of conservation prioritization compared with CLIP 3.0 0 top 3 priorities under 1.0 m SLR. Percentages indicate what percent of conservation priorities identified by Zonation falls within CLIP

Zonation Conservation Priorities	Within CLIP (Acres)	Not Within CLIP (Acres)	Percentage of Zonation Priorities in CLIP
2%	5,238	43	99.2%
2%-5%	7,365	557	93.0%
5%-10%	11,983	1,220	90.8%
10%-25%	35,508	4,101	89.7%
25%-50%	58,715	7,301	88.9%
50%-75%	57,337	8,678	86.9%
75%-100%	26,911	39,104	40.8%

Table 2-13 Statistics of conservation prioritization compared with CLIP 3.0 0 top 3 priorities under 2.5 m SLR. Percentages indicate what percent of conservation priorities identified by Zonation falls within CLIP

Zonation Conservation Priorities	Within CLIP (Acres)	Not Within CLIP (Acres)	Percentage of Zonation Priorities in CLIP
2%	5,027	255	95.2%
2%-5%	7,265	658	91.7%
5%-10%	12,429	776	94.1%
10%-25%	35,047	4,569	88.5%
25%-50%	56,956	9,071	86.3%
50%-75%	59,425	6,602	90.0%
75%-100%	26,908	39,070	40.8%

Table 2-14 Statistics of conservation prioritization with development projection under 0.5 m SLR

0.5 m SLR Zonation Conservation Priorities	Projected Development Within Conservation Priorities (Acres)	Projected Development not Within Conservation Priorities (Acres)	Percentage of Zonation Priorities in Development
2%	327	4,954	6.20%
2%-5%	474	7,448	5.98%
5%-10%	827	12,376	6.26%
10%-25%	3,470	36,139	8.76%
25%-50%	13,417	52,598	20.32%
50%-75%	13,337	52,678	20.20%
75%-100%	10,801	55,213	16.36%

Table 2-15 Statistics of conservation prioritization with development projection under 1.0 m SLR

1.0 m SLR Zonation Conservation Priorities	Projected Development Within Conservation Priorities (Acres)	Projected Development not Within Conservation Priorities (Acres)	Percentage of Zonation Priorities in Development
2%	322	4,959	6.11%
2%-5%	452	7,470	5.71%
5%-10%	814	12,389	6.16%
10%-25%	4,209	35,400	10.63%
25%-50%	12,380	53,636	18.75%
50%-75%	13,677	52,338	20.72%
75%-100%	10,800	55,215	16.36%

Table 2-16 Statistics of conservation prioritization with development projection under 2.5 m SLR

2.5 m SLR Zonation Conservation Priorities	Projected Development Within Conservation Priorities (Acres)	Projected Development not Within Conservation Priorities (Acres)	Percentage of Zonation Priorities in Development
2%	288	4,993	5.46%
2%-5%	397	7,525	5.01%
5%-10%	751	12,452	5.69%
10%-25%	5,232	34,377	13.21%
25%-50%	11,403	54,612	17.27%
50%-75%	13,581	52,434	20.57%
75%-100%	11,002	55,013	16.67%

CHAPTER 3
INCORPORATING ECONOMIC CONCEPTS INTO SEA LEVEL RISE ADAPTIVE CONSERVATION PLANNING AND RESERVE DESIGN

3.1 Economic Concepts in Conservation Planning and Reserve Design

Although the biological aspects of conservation planning have advanced rapidly, little attention has been paid to the economic aspects of conservation planning (Naidoo & Ricketts, 2006). There are several reasons for this. Firstly, the spatially-explicit economic information needed in conservation planning and reserve design is usually not readily available to conservation planners (Naidoo et al., 2006). Secondly, quantifying the economic value of ecosystem services provided by natural systems is inherently difficult because of the complexity of ecosystem functions. Thirdly, the field of conservation planning and reserve design was developed by conservation biologists and most biologists were not trained to consider cost analysis in the planning process (Naidoo et al., 2006).

Integrating economic considerations into conservation planning and reserve design has become a hot topic in the literature of conservation biology (Naidoo et al., 2006; Arponen et al., 2010). More and more studies have demonstrated that it is important and necessary to apply economic considerations to tools in conservation decision making (Naidoo et al., 2006; Watzold et al., 2006). For example, a study on European wetlands conservation found that ignoring land rent adjustments in the planning process could lead to highly cost-ineffective solutions in reserve selection (Jantke, 2011). Naidoo & Ricketts (2006) conducted a spatial evaluation of the costs and benefits of conservation in the Mbaracayu Biosphere Reserve in Paraguay and found that there was a high degree of spatial variability in both costs and benefits over this reserve thus understanding the trade-offs between conservation and economic development can powerfully inform conservation decision-making.

Though considering economic factors is important in conservation decision-making, there is concern about economic considerations overriding biological considerations in conservation planning. Protecting biodiversity should always be the first and foremost target for conservation planning (Naidoo et al., 2006), and focusing too narrowly on economic factors in conservation planning may lead to the opportunistic selection of conservation areas that often could not protect biodiversity efficiently (Arponen et al., 2010). Economic facts are an important component in conservation decision-making, but understanding how much weight should be given to economic considerations is crucial (Arponen et al., 2010).

As sea level rises, coastal habitats will change in response to this rise. Economic factors including costs and benefits of conservation associated with protecting coastal habitats will change accordingly. Economic cost analysis will be useful to coastal reserve managers and other conservation decision-makers as they will need the spatially-explicit information about how their current conservation plans should be revised (e.g., enlarge the existing reserves, create new conservation areas, and/or establish ecological corridors) to facilitate biodiversity adaptation to SLR while minimizing the costs spent to achieve the targeted conservation goals.

The need for conservation efficiency in SLR adaptive conservation planning from natural resource managers provides a good opportunity for conservation planners to integrate economic concepts into SLR adaptation planning process. Fortunately, spatial economics, which deals with the allocation of scarce resources over space, and the location of economic activity, have been increasingly incorporated into conservation planning and reserve design (Naidoo et al., 2006). To provide conservation decision-makers explicit information about economic factors associated with conservation, this chapter quantifies costs (dollar cost of land protection) and benefits (economic value of ecosystem services) of conservation and integrates these economic concepts into SLR adaptive conservation prioritization in the Matanzas study area.

3.2 Methods

3.2.1 Estimating Conservation Costs

"All conservation interventions have associated costs, which cover everything that must be given up to implement the intervention." (Naidoo et al., 2006). Naidoo (2006) summarized five different types of conservation costs including acquisition costs (costs of acquiring land properties), management costs (costs of managing a conservation program), transaction costs (costs of negotiating an economic exchange), damage costs (costs of damage due to conservation activities, e.g. damages to livestock because of wildlife attack) and opportunity costs (costs of foregone opportunities, e.g. economic loss of protecting land for conservation purposes instead of growing agricultural crops). A full accounting of conservation costs would include all five, but this is likely to be difficult or impossible in most planning exercises and typical studies would select the largest component of conservation costs as a surrogate for all costs (Adams et al., 2010). Usually, the cost of creating conservation areas is dominated by land acquisition and long-term management costs, and it has been estimated that land acquisition cost is likely to exceed management costs by large factors (Balmford et al., 2003; Adams et al., 2010).

Land acquisition costs are costs associated with buying land property rights and placing land in protected status. In developed countries like the United States, such costs can be directly estimated by land prices or assessed land values (Ando et al., 1998; Polasky et al., 2001; Naidoo et al., 2006). In Florida, the Florida Department of Revenue (FDOR) publishes vector digital parcel data that contain land values from FDOR's tax database for each parcel. To estimate the conservation land acquisition costs in the Matanzas study area, parcel data from FDOR was clipped into the study area and converted into raster data showing land value per acre based on the data field land value.

3.2.2 Integrating Conservation Costs in SLR Adaptive Conservation Planning

Costs can be included in conservation priority process in Zonation to achieve conservation efficiency (Moilanen et al., 2012). This is achieved by inputting a grid layer containing conservation cost information in the analysis process. In Zonation, the costs do not necessarily have to be measured in terms of money. Other measures of economic losses such as intensity of harvesting can also be used as surrogates for the cost layer (Moilanen et al., 2012).

The Zonation algorithm runs by selecting cells that have a high biodiversity conservation value/cost ratio. For the Matanzas study area, this cost layer is the estimated land acquisition cost based on parcel data. The current land acquisition costs were used to represent land acquisition costs under the 0.5 m, 1.0 m and 2.5 m SLR scenarios because the selection of conservation priorities is based on the conservation value/cost ratio and the relative conservation value/cost ratio for each cell will not change if we assume that land value for each parcel will increase at the same rate in the future though this might not be the case in various locations. For example, land adjacent to the coastline may become relatively cheaper over time as SLR progresses, but this would be difficult to quantify and include in this type of analysis. Figure 3-1 shows the integrated modeling framework used to identify SLR adaptive conservation priorities when including conservation costs in the analysis.

Inputs for running Zonation to include consideration of conservation costs in this research include:

- A set of biodiversity feature grid layers the same as those used in Chapter 2.
- A cost grid layer represents land acquisition costs based on assessed value in parcel data from FDOR.
- A biodiversity feature list file the same as the biodiversity feature list file used in Chapter 2.
- A run settings file that contains all basic Zonation settings but includes costs in the analysis and excludes areas that have no development value (Appendix E).

3.2.3 Estimating Conservation Benefits

Despite the importance of ecosystems that yield a flow of services to human communities, this natural capital is poorly understood and scarcely monitored (Tallis et al., 2008). To help distribute and share the information of ecosystem services valuation studies, the GecoServ database (http://www.gecoserv.org/) was developed to quantify ecosystem service valuation in the Gulf of Mexico region. The database summarizes ecosystem services valuation by each category based on extensive literature review. For this study, the database was used to calculate changing ecosystem services valuation due to SLR in the Matanzas study area. The habitat categories were based on SLAMM outputs and converted to relevant categories in the GecoServ database. The value of each ecosystem service provided by each ecosystem type was calculated by using the average value of that service in the database and only those values that substantiated by research conducted in the United States after 2005 (Table 3-1). The results show that beach, mangrove and freshwater wetlands could provide approximately 79,000, 50,000 and 25,000 U.S. dollars per acre per year respectively, which are the most productive habitats in the study area.

3.3 Results

3.3.1 Conservation Costs

Figure 3-2 presents the acquisition cost of conservation in the Matanzas study area based on parcel data from FDOR. The map shows that Land value varies significantly with high values in the northeast and southern portions of the study area, where the cities of St.

Augustine and Palm Coast are located.

3.3.2 Conservation Prioritization Considering Economic Costs

Figure 3-3 presents conservation prioritization when considering costs in the planning process under current conditions, 0.5 m, 1.0 m and 2.5 m SLR scenarios in the Matanzas study area. Results show that top priorities are located in the southwestern portion of the study area. This is due to the low costs of land value in these areas based on parcel data. Figure 3-4 shows the comparison of top 10% conservation priorities under the four scenarios. The green color represents top 10% conservation priorities identified in all of the four SLR scenarios. The cause of this change is that the biological value of each cell will change as habitat changes due to SLR.

When compared to conservation prioritization done in Chapter 2 that did not consider land costs in the Zonation analysis process, the conservation prioritization incorporating costs in this chapter changes priority results significantly. Figure 3-5 shows the differences of top 10% conservation priorities identified in these scenarios with and without consideration of economic costs. The comparison demonstrates that some areas with high conservation values might be identified as lower priorities when integrating economic costs in the planning process and some areas with low conservation values might be identified as high priorities when considering costs in the planning process. This change is due to the change of the algorithm used in Zonation that the selection of conservation priorities is based on cells which have a high biodiversity conservation value/cost ratio instead of purely the biodiversity conservation value of the cell.

3.3.3 Conservation Benefits

Table 3-2 shows the changing areas of each habitat type due to SLR in the Matanzas study area. The areas were calculated based on SLAMM results and grouped into GecoServ categories including freshwater wetlands, saltwater wetlands, mangrove, beach and marine/open water. Ecosystem service values for uplands were not accounted for in this analysis because the GecoServ database does not have valuation for uplands. However, to include ecosystem service valuation for uplands in the analysis, a uniform value under the lowest ecosystem value of saltwater wetlands which is $11,585 per acre per year was used to substitute the actual ecosystem value provided by uplands.

Based on this equation: Ecosystem Services Valuation = Acres of habitat * Average Value (U.S. $ per acre per year from GecoServ), ecosystem services valuation under each SLR scenarios was calculated and a dollar amount value X was given to uplands to be included in the analysis (Table 3-3).Based on SLAMM results and ecosystem service types and valuations from GecoServ, the results show that the value of ecosystem services provided by freshwater wetlands, mangrove and uplands will decrease as sea level rises but the values of ecosystem services provided by saltwater marsh, beach and marine/open water will increase as sea level rises. This is due to the loss of ecosystem types including freshwater wetlands, mangrove and uplands but the gain of ecosystem types including saltwater marsh, beach and open water as sea level rises based on SLAMM results. The total ecosystem service value will decrease when sea level rises based on this analysis, and this is largely because the loss of ecosystem services provided by freshwater wetlands and uplands cannot be offset by the gain of ecosystem services provided by saltwater marsh and open water.

3.4 Discussion

Conservation is both costly and beneficial. Spatially-explicit information about costs and benefits of conservation is useful to conservation decision-makers as they will need this information to allocate funds and resources efficiently (Naidoo et al., 2006). The conservation costs and benefits analysis demonstrate variation in the Matanzas study area. Understanding the economic side of SLR adaptive conservation planning will help GTMNERR manager and other coastal resources managers to make informed decisions about where and how to allocate resources more wisely to facilitate biodiversity adaptation to SLR.

As shown, costs of conservation can be incorporated into the conservation prioritization process using Zonation. When considering costs in the planning process, the conservation priority outcomes changed a lot compared to conservation outcomes without consideration of costs. The conservation prioritization that integrates costs in the planning process will provide important information for coastal reserve managers and other decision-makers about where to acquire additional conservation lands to facilitate biodiversity adaptation to SLR while minimizing costs associated with conservation activities. For example, we might have two areas that could be acquired as important additions to existing conservation and the conservation prioritization will provide information for reserve managers to decide which option is better from a cost/benefit perspective.

The conservation benefits analysis shows that the value of ecosystem services provided by freshwater wetlands will decrease significantly as sea level rises. As a high value ecosystem type, strategies such as wetland rehydration, dispersed water storage, and wetland mitigation banks could be taken to restore freshwater wetlands in the study area to offset these losses. The benefits of conservation analysis could also help inform conservation decision makers achieve long-term conservation goals. However, this information is likely insufficient to motivate conservation because the financial mechanism and institutions have not yet been established to fully include them in the market though this has been discussed a lot (Naidoo et al., 2006). The reality is that in most cases, landowners could not get compensation for bearing the costs of protecting the areas that provide these services (Pagiola et al., 2002; Naidoo et al., 2006). Without an established financial mechanism for including ecosystem services in the market, the value of ecosystem services provided by natural habitats will continue to be ignored by planners that make conservation and land-use decisions.

In the analysis of both costs and benefits of conservation, there were shortcomings and limitations in this research. Because of timing and data limitations, only land acquisition costs were calculated. Other costs of conservation including opportunity costs, management costs, transaction costs and damage costs were not included in this analysis. Though conservation land acquisition costs are likely to dominate costs of conservation (Balmford et al., 2003; Adams et al., 2010), ignoring other costs in conservation might risk overlooking some important parts of costs associated with conservation. In addition, the analysis of conservation benefits in this research was based on SLAMM results and ecosystem services valuation from the GecoServ database. The SLAMM results show that mangrove will decrease significantly as sea level rises. However, mangrove could also increase due to potential temperature increases which are not accounted for in the SLAMM modeling. Therefore, the total area of mangrove may increase significantly at the expense of saltmarsh in the future. In addition, not all services provided by each type of ecosystem were accounted for and only those services that could be quantified were included in this analysis. Therefore, the valuation estimates for ecosystem services might be significantly lower than the actual ecosystem services provided by natural communities in the Matanzas study area. Last, using the estimates of Gulf of Mexico ecosystem service benefits in an Atlantic coastal ecosystem has potential limitations. The valuation of ecosystem services in the Gulf of Mexico region could be different from the valuation of ecosystem services in an Atlantic coastal region (e.g., coastal wetlands in the Gulf of Mexico region may provide higher ecosystem service value as this region is more likely to be hit by hurricanes).

The economic analysis of conservation could inform conservation and land use to achieve the goal of conservation efficiency. The conservation prioritization that integrates costs in the planning process was based on a favorable biodiversity conservation/cost ratio. Understanding this information is useful to conservation decision making especially when the two are aligned and we need to select priorities with a limited budget. However, the economic considerations of conservation can never override the ethical arguments for conservation (Naidoo et al., 2006). Protecting biodiversity should always be the first and foremost target for conservation, not a favorable biodiversity conservation/cost ratio. In this research, incorporating cost into conservation prioritization via Zonation might complicate consideration of cost, which could be done in a simpler and straightforward way by looking at land costs post facto. The conservation costs analysis is most useful when we need to select an area for conservation while the potential two or more areas have the same biological priority as identified in Chapter 2. In this case, selecting the area with the lowest acquisition cost would be the optimal solution.

Table 3-1 Average value in 2012 U.S. $ per acre per year of each service provided by coastal habitats from GecoServ

<table>
<tr><th>SLAMM Category</th><th>GecoServ Category</th><th>Service</th><th>Average Value</th></tr>
<tr><td rowspan="4">Swamp</td><td rowspan="10">Freshwater wetland</td><td>Habitat</td><td>$1,728.56</td></tr>
<tr><td>Disturbance regulation</td><td>$2,248.64</td></tr>
<tr><td>Aesthetics, Recreation</td><td>$1,986.70</td></tr>
<tr><td>Waste regulation</td><td>$3,451.16</td></tr>
<tr><td rowspan="3">Cypress swamp</td><td>Raw materials</td><td>$1428.37</td></tr>
<tr><td>Water supply</td><td>$5,138.88</td></tr>
<tr><td>Water regulation</td><td>$5,534.86</td></tr>
<tr><td rowspan="4">Inland freshwater marsh</td><td>Climate regulation</td><td>$264.57</td></tr>
<tr><td>Nutrient Cycling</td><td>$273.05</td></tr>
<tr><td>Food</td><td>$2,521.10</td></tr>
<tr><td rowspan="8">Freshwater wetland</td><td>Medicinal Resources</td><td>$216.92</td></tr>
<tr><td rowspan="3">Transitional saltmarsh</td><td>Subtotal</td><td>$25,074.49</td></tr>
<tr><td>Disturbance regulation</td><td>$1,361.80</td></tr>
<tr><td>Biological control</td><td>$121.81</td></tr>
<tr><td>Regularly flooded marsh</td><td>Waste regulation</td><td>$4,439.30</td></tr>
<tr><td>Tidal flat</td><td>Water supply</td><td>$65.36</td></tr>
<tr><td rowspan="2">Irregularly flooded marsh</td><td>Water regulation</td><td>$4,765.07</td></tr>
<tr><td>Nutrient cycling</td><td>$11.17</td></tr>
</table>

Continued

SLAMM Category	GecoServ Category	Service	Average Value
Vegetated tidal flat	Saltwater marsh	Gas regulation	$520.03
		Food	$99.76
		Cultural, spiritual and historic Subtotal	$125.99
		Aesthetics and recreation	$75.68
		Subtotal	$11,585.97
		Habitat	$35.90
		Disturbance Regulation	$1,261.03
		Waste Regulation	$1,921.49
		Raw Materials	$15,424.93
Mangrove	Mangrove	Nutrient regulation	$17.81
		Gas regulation	$391.34
		Food	$9,556.05
		Bequest, Existence, Option	$7,030.76
		Recreation	$15,348.85
		Subtotal	$50,988.14
Estuarine Beach	Beach	Disturbance Regulation	$28,520.84
		Erosion Control/Soil Retention	$33,589.64
Ocean Beach		Cultural, Spiritual & Historic	$23.48
		Aesthetics, Recreation	$17,120.60
		Subtotal	$79,254.56

Table 3-2 Habitat change under the 0.5 m, 1.0 m and 2.5 m SLR scenarios

Habitat type (Acres)	Initial condition (Acres)	0.5 m SLR (Acres)	1.0 m SLR (Acres)	2.5 m SLR
Freshwater wetlands	76,471	75,354	73,552	69,319
Salt water wetlands	12,888	19,216	18,280	21,249
Mangrove	387	287	222	1
Beach	543	556	668	1,276
Open Water	12,703	13,859	20,149	33,995
Upland	120,653	116,268	114,731	108,578

Table 3-3 Ecosystem services valuation under the 0.5 m, 1.0 m and 2.5 m SLR scenarios

Habitat type	Initial condition (U.S. $)	0.5 m SLR (U.S. $)	1.0 m SLR (U.S. $)	2.5 m SLR (U.S. $)
Freshwater wetlands	1,917,489,540	1,889,464,031	1,844,299,103	1,738,142,229
Saltwater wetlands	149,323,332	222,642,589	211,801,134	246,200,224
Mangrove	19,743,119	14,671,838	11,360,668	53,028
Beach	43,066,931	44,079,011	52,952,352	101,140,713
Marine/open water	1,497,682	16,339,317	23,754,378	40,078,511
Upland	1,397,882,038	1,347,077,560	1,329,269,924	1,257,981,451
Total	3,527,504,960	3,534,274,346	3,473,437,559	3,383,596,156

CHAPTER 4
THE VALUE OF DATA ACCURACY AND MODEL SOPHISTICATION IN SEA LEVEL RISE ADAPTIVE CONSERVATION PRIORITIZATION

4.1 Uncertainties in Sea Level Rise Adaptive Conservation Planning

4.1.1 Uncertainty in Choosing Elevation Data and Coastal Impact Models

It is important to accurately identify and delineate conservation priorities because conservation and land-use decision makers will need this critical information to guide them. However, SLR and other effects of global climate change produce a decision making environment marked by high uncertainty (Noss, 2011). Uncertainty exists around the appropriate choice of impact assessment methods with different models and parameterizations potentially producing different results (McLeod et al., 2010). Approaches for modeling the response of coastal ecosystems to SLR vary in their level of intricacy and data requirements, from simplistic "bathtub" applications of SLR projections, to more complex models that account for hydrological processes, ecological feedbacks and anthropogenic barriers to habitat transition dynamics (McLeod et al., 2010).

High resolution elevation data (Lidar; Light Detection and Ranging) and sophisticated coastal impact models (SLAMM) are preferably used in SLR adaptation planning to predict habitat change and provide foundations for identifying conservation priorities in response to SLR and land use change (Runting et al., 2013). However, we know little about whether the improved model-dataset combination would actually lead to more efficient conservation outcomes and to what extent this model-dataset combination could improve SLR adaptive conservation outcomes. This is important to know because high resolution data is likely not to be available in some data-poor coastal areas and running more complicated models is relatively time consuming, expensive and needs additional expert knowledge.

However, a recent study that examines the cost-effectiveness of different model-dataset combinationsin coastal South East Queensland, Australia showed that it is considerably more cost-effective to use a process-based model and high-resolution elevation data in SLR adaptive conservation designs, even if this requires a substantial proportion of the project budget to be spent. The less accurate model and dataset failed to identify areas of high conservation value, reducing the cost effectiveness of the resultant conservation plan (Runting et al., 2013). As Runting (2013) pointed out that a future research priority is to quantify how the project findings vary among different regions because the findings may not hold in all situations, particularly those areas with consistently low topographic relief or different economic context. This research investigates the differences among conservation prioritizations produced by different model-dataset combinations in a consistently low topographic relief area in Florida.

4.1.2 Uncertainty in Elevation Dataset Vertical Accuracy

Sources of possible uncertainty in DEM datasets include data errors such as the age of data, measurement errors such as positional inaccuracy and processing errors such as interpolation errors (Burrough,1986; Wise, 1988). According to United States Geological Survey (USGS), the accuracy of the National Elevation Dataset (NED) varies spatially because of the variable quality of the source digital elevation models. USGS maps follow the national mapping standards for vertical and horizontal precision. Vertically, the standard is that 90 percent of the well-defined points along a contour must be within one-half the contour interval above or below the stated elevation of the contour (Titus & Cacela, 2008). The overall absolute vertical accuracy which expressed as the root mean square error (RMSE) is usually less than 2.44 meters for NED. Horizontally, the horizontal standard is that 90 percent of the points should be within one fiftieth of an inch (Titus & Cacela, 2008). According to Titus and Cacela (2008), the Lidar data sources generally have vertical precision on the order of 10–30 centimeters and horizontal error of less than 1.0 meter.

The accuracy of elevation data is critically important for predicting SLR impacts on coastal habitats. However, the inherent uncertainty in elevation data could limit this accuracy and thus limit the accuracy of predictions and the usefulness of the elevation data (Gesch, 2009). In most previous studies using elevation data to estimate potential inundation from SLR, the uncertainty of elevation data has not been taken into account in the analysis (Small & Nicholls, 2003; Ericson 2006; Gesch, 2009). The reliability of the estimates of potential inundated areas is questionable because the results may not represent the whole picture of SLR impact due to the inherent uncertainty in DEMs (Gesch, 2009). Gesch (2009) compared different elevation datasets with varying resolution and accuracy in delineating lands vulnerable to SLR and found that high resolution Lidar data could not only provide a much more detailed delineation of inundation areas but also improve the certainty of the delineation of lands that will be impacted by SLR. To include uncertainty of DEMs in the analysis, this study includes a quantification of the vertical accuracy of NED and Lidar that have been used in identification of SLR adaptive conservation priorities in the Matanzas study area.

4.2 Methods

4.2.1 Sea Level Rise Scenario

To compare conservation outcomes of different model-dataset combinations, a 1.0 meter SLR projection by 2100 is used to predict the impacts of SLR on coastal habitats in the Matanzas study area. I chose the 1.0 m SLR scenario for several reasons. Firstly, the 1.0 m SLR projection by the end of the 21st century is the “likely” worst case scenario according to the 5th assessment report (AR5) of IPCC (Church et al., 2013). Secondly, it is likely not necessary to include different SLR scenarios in this analysis as the aim of this study is to examine conservation outcomes of different model dataset combinations though this analysis might be useful in the future because differences in predictions and identified priority areas might be significantly different for different SLR levels. Thirdly, the 1.0 m SLR projection has been used in a previous similar study to compare inundation lands vulnerable to SLR with different elevation datasets (Gesch, 2009).

4.2.2 Model-dataset Combinations

Using different coastal impact models and different elevation datasets can result in different conservation outcomes. To compare the conservation prioritization with different model-dataset combinations, two coastal impact models (SLAMM and Bathtub) and two elevation datasets (Lidar and NED) were selected for comparison. The SLAMM is a “complex” coastal impact model that simulates the geomorphological processes that result in coastal wetland conversions and shoreline modifications during long-term SLR (Clough et al., 2010). The bathtub model is a “simplistic” model that works by identifying lands under a given contour as inundation areas. In addition to the coastal impact models, the quality of elevation data as input data is another important factor that determines conservation outcomes in response to SLR. Recent years, high quality elevation data derived from Lidar has become available and this “fine” resolution dataset is highly suitable for detailed study related to SLR (Gesch, 2009). For this study, “fine” data was a composite from St. Johns County Lidar (2008), Palm Coast Lidar (2008), and FWC-FWRI (2009), and the “coarse” data was NED from USGS. The two models and datasets can form four different model-dataset combinations including SLAMM-Lidar, SLAMM-NED, Bathtub-Lidar and Bathtub-NED. For this study, the SLAMM-NED combination was not included in the analysis due to the lack of availability of this model-data combination from the Matanzas study. Table 4-1 shows the model-dataset combinations used in identifying SLR adaptive conservation prioritization in the Matanzas study area.

4.2.3 Comparison of Conservation Prioritization with Different Model-Dataset Combinations

The overall methodology used in this research to compare the conservation prioritization outcomes of different model-dataset combinations is shown in the below diagram (Figure 4-1). Based on the three model-dataset combinations (SLAMM-Lidar, Bathtub-Lidar and Bathtub-NED), three different land use datasets are generated as the key input to run species habitat models for selected focal species in the Matanzas study area. The land use grids were created based on combinations of the current land use dataset from St. Johns River Water Management District (SJRWMD) and the changed land use cells due to SLR. The SLAMM codes were crosswalked to FLUCCS codes (Appendix F) and the inundation areas identified by the bathtub model were given a value of 5000 as open water. Using the land use grids as the primary input for species habitat models, species habitat scenarios were identified for 5 selected focal species in the Matanzas study area to demonstrate the impact of model-dataset combinations on species habitat identification. The selected 5 species include black rail, limpkin, painted bunting American oystercatcher and Worthington’s marsh wren, in which all of their habitats are considered to be significantly affected by SLR.

Outputs of the species habitat models will be used as primary inputs for running Zonation software to identify conservation priorities adaptive to 1.0 m SLR. The SLAMM-Lidar combination is set to be the benchmark scenario that the other two model-dataset combinations are compared in order to measure the differences of conservation outcomes with different model-dataset combinations. Specifically, inputs for running Zonation in this study to compare conservation outcomes with different model-dataset combinations include:

- A set of biodiversity feature grid layers which are the selected species habitat grids derived from species habitat models.
- A biodiversity feature list file which is the same as the biodiversity feature list file used in Chapter 2.
- A run settings file that contains all basic Zonation settings which is the same as the run setting file used in Chapter 2.

4.2.4 Application of Vertical Accuracy as a Measure of DEM Uncertainty

Linear error is the metric used by the National Standard for Spatial Data Accuracy (NSSDA) (Gesch, 2009; Federal Geographic Data Committee, 1998). Gesch (2009) has calculated linear error with a 95% confidence level and used it to measure uncertainty of inundation areas to 1.0 m SLR in eastern North Carolina. The linear error for each elevation dataset with a 95% confidence level (CI) was calculated based on the method described in Maune, Maitra, and McKay (2007) with the below expression:

Linear error at 95% confidence = 1.96 * RMSE

Table 4-2 shows characteristics of the two elevation datasets used in this study. The linear error at 95% CI for Lidar elevation is less than 0.3 m, which is much less than the linear error for NED. To show a spatial representation of the uncertainty of DEMs in SLR impacts, additional areas that are likely to be inundated by 1.0 m SLR are delineated by adding the linear error at 95% CI.

4.3 Results and Discussion

4.3.1 Sea Level Rise Vulnerability Assessment

To demonstrate the impact of SLR on coastal areas with different elevation datasets, two vulnerability maps showing inundation zones from a 1.0 m SLR were created using a bathtub model. Land cells that are at or below 1.0 m SLR and are connected hydrologically to the ocean were identified as "flooded" areas (Poulter &Halpin, 2007). The dark blue color shows potential inudation areas from 1.0 m SLR and the light blue shows uncertainty associated with the elevation datasets (Figure 4-2).

The area of inundation from 1.0 m SLR was calculated from the two different elevation datasets and the uncertainties associated with elevation datasets were also caculated (Table 4-3).The results show that inundation areas delineated from the Lidar-based elevation dataset aremuch less than the inundation areas delineated from the NED dataset. When elevation uncertainty was included in the delineation process, more than 50% additional lands was identified as being subject to a 1.0 m SLR based on the NED dataset but only approximately 10% additional lands was identified as being subjected to a 1.0 m SLR based on the Lidar elevation dataset.

A large area of additional lands will be projected as inundated by a 1.0 m SLR based on the NED dataset because lots of additional inundation lands will be identified as being subject to the 2.21 m vertical accuracy. Much less additional indunation lands are idenfitifed based on the Lidar elevation dataset because the linear error for Lidar is less than 0.3 m. Inundation areas delineated in the Lidar elevation dataset either with or without consideration of elevation uncertainty are much less than that in the NED dataset because the higer resolution elevation dataset (Lidar-based) could aviod delineation of extra inundation areas to SLR while these areas could be idenfied as being subject to SLR in the lower resolution elevation dataset (NED). This comparison demonstrates the advantages of Lidar elevation dataset in delineation of inundation areas being subject to SLR even when using simplistic bathtub models. The accuracy of delineations is especially important for coastal decision-makers in areas with very low topographic relief such as Florida.

4.3.2 Land Cover Impacts with Different Model-dataset Combinations

Conservation planners and reserve managers responsible for mitigating SLR effects will need accurate estimation of land cover types to be affected by potential SLR (Gesch, 2009). To demonstrate SLR impacts on land cover types with different model-dataset combinations, land cover maps showing water and wetland distribution as sea level rises of 1.0 m were created using all three methods (Figure 4-3). Compared with current land cover, wetland areas at the 1.0 m SLR scenario are likely to decrease with the bathtub-NED and bathtub-Lidar combinations but increase with the SLAMM-Lidar combinations. This is because the bathtub model can only indicate potential inundation of existing wetlands and other terrestrial landcover and not the potential conversion of uplands to wetlands that will often occur during SLR. Compared with current land cover, water areas at the 1.0 m SLR scenario were projected to increase significantly with all the three model-dataset combinations (Table 4-4). The increase of water is due to large areas being inundated by the rising sea water and low topographic relief, especially using the bathtub model. The results show that using different model-dataset combinations in predicting land cover change under SLR will clearly lead to differences in the estimated areas of projected future land cover types, and therefore habitat delineation for focal species could also be significantly affected.

4.3.3 Species Habitat with Different Model-dataset Combinations

Using species habitat models from Dr. Tom Hoctor at University of Florida's Center for Landscape Conservation Planning and FWC, four species habitat maps including current species habitat and species habitat under the three different model-dataset combinations were created for the5 selected species (Figure 4-4 to Figure 4-8). Statistics of the areas of each species habitat show that using different model-dataset combinations in predicting species habitat under SLR will clearly lead to the differences in the estimated areas of species habitat (Table 4-5).

Black rails are secretive marsh birds and this bird is usually found in high saltmarsh and some freshwater marshes where soils are inundated or have only extermely shallow water (Cox & Kautz, 2000). The habitat model identifies the higher saltmarsh that are within 300 meters of uplands and more than 100 meters further away from open water that black rail appear to prefer (Hoctor, 2011). Compared to black rail's current habitat, all the three model-dataset combinations identify much less habitat under the 1.0 m SLR scenario. This is because higher saltmarsh further away from water is decreasing since open water increases as SLR. Compared to black rail habitat identified in the SLAMM-Lidar combination, habitat identified using the bathtub model is larger. This is because saltmarshes that could meet the modeling requirements (within 300 meters of uplands and more than 100 meters away from open water) will decrease based on SLAMM-Lidar combination.

Limpkins are found in a variety of freshwater wetlands and water bodies in Florida. They usually occur in freshwater wetlands where apple snails and other snails appear. They are also found near lake, river and stream edges, wetlands associated with open water and herbaceous freshwater wetlands (Cox, 1994). The habitat model idenfifies all freshwater wetlands adjacent to appropriate open water and all freshwater herbaceous wetlands as limpkin habitat (Hoctor, 2011). Compared to limpkin's current habitat, all three model-dataset combinations identify less habitat under the 1.0 m SLR scenario. This is because freshwater will decrease as sea level rises in all the three model-dataset combinations. The SLAMM-Lidar combination could identify more limpkin habitat than the other two combinations because the SLAMM model takes wetland conversions into account while the bathtub model does not, although SLAMM may still under-represent conversion of uplands to freshwater marsh. The bathtub-NED combination identifies the least limpkin habitat because a large number of freshwater wetlands are likely to be inundated in this scenario.

Painted bunting is usually found in xeric oak scrub, shrub and brushlands in coastal areas. This species breeds in maritime hammocks or scrub byland cover (Lowther et al., 1999). The habitat model identifies all xeric oak scrub, shrub and brushland, and then hardwood hammocks and forest and cabbage palm-live oak hammocks that were located on "extremely well", "well", or moderately-well drained soils within 60 m of the xeric oak and brushland classes as potential habitat (Hoctor, 2011). Compared to painted bunting's current habitat, all the three model-dataset combinations identifies less habitat. This is because the appropriate habitat for this species will decrease as sea level rises of 1.0 m in the future since it is exclusively dependent on drier upland sites. The SLAMM-Lidar combination identifies more painted bunting habitat than the other two combinations because it provides a higher resolution depiction of upland loss than bathtub models.

Habitat for American oystercatcher include shell bars, spoil islands, coastal beaches, mudflats and salt marshes and this species prefers large sand areas with sparse vegetation for nesting (Lauro & Burger, 1989). The habitat model identifies non-vegetated wetlands, tidal flats, shorelines, oyster bars, beaches other than swimming beaches and exposed rock with marsh grasses where it has been found to occur as appropriate American oystercatcher habitat (Hoctor, 2011). Compared to American oystercatcher's current habitat, the SLAMM-Lidar combination scenario identifies much more additional habitat but the bathtub-NED and bathtub-Lidar combinations identifies no habitat for this species. This is because the appropriate American oystercatcher habitat which includes non-vegetated wetlands, tidal flats and other habitats is projected as inundated by the rising sea water based on simplistic bathtub models. While using SLAMM, appropriate American oystercatcher habitat can be identified because this coastal impact model recongnizes wetland conversions and less inundation as sea level rises.

Habitat of Marian's marsh wren includes tidal marshes, especially salt marshes adjacent to tidal creeks with little or no mangrove encroachment (Rodgers et al., 1996). The habitat model identifies saltwater marsh within 150 m of tidal creeks as primary habitat and saltwater marsh over 150 m from tidal creeks as secondary habitat for Marian's marsh wren (Hoctor, 2011). Compared to Marian's marsh wren's current habitat, the SLAMM-Lidar combination scenario identifies 18.27% less habitat and the bathtub-NED and bathtub-Lidar combinations identifies no habitat for this species. This is because saltwater marshes within/ over 150 m of tidal creeks will be all inundated by the 1.0 m SLR based on the bathtub model but most of this habitat will remain unaffected and some new habitat will be gained based on the SLAMM analysis.

The selected species habitat analysis demonstrates that using different model-dataset combinations could clearly lead to different outcomes of species habitat in response to SLR. Species habitat identified based on SLAMM could be much more than species habitat identified based on the bathtub model for estuarine species such as American oystercater and Marian's marsh wren. The results show that appropriate selection of coastal impact models and elevation dataset is very important in identifying species habitat under SLR and using the less sophisticated coastal impact model and less accurate elevation dataset could fail to identify a large number of appropriate habitat for some species.

4.3.4 Conservation Prioritization with Different Model-dataset Combinations

Using the 5 selected species habitat grids as input "biodiversity features" for Zonation, four conservation prioritization scenarios including current conservation prioritization and conservation prioritization under 1.0 m SLR were identified based on the different model-dataset combinations (Figure 4-9). Using conservation outcomes of the SLAMM-Lidar combination as benchmark, a large proportion of the top conservation priorities identified by the SLAMM-Lidar combination is not identified by the bathtub-NED and bathtub-Lidar combinations. Figure 4-10, Figure 4-11 and Figure 4-12 show the differences of top 10% conservation priorities in these scenarios. Compared to the top 10% conservation priorities identified via the SLAMM-Lidar combination, the bathtub-NED combination could miss 17,815 acres of land and the bathtub-Lidar combination could miss 16,433 acres of lands not identified as their top 10% priorities respectively. The results indicate that different model-dataset combinations could lead to significant differences in conservation prioritization outcomes and the less accurate model and dataset could fail to identify areas of high conservation value for adaptation to SLR. The spatially-explict analysis in this research could inform coastal reserve managers and other decision makers about the importance of sophisticated coastal impact model and high-resolution elevation data in the identification of conservation priorities in response to SLR.

4.4 Conclusions

Inherent uncertainty exists in elevation datasets, and uncertainty exists around the appropriate choice of coastal impact models and elevation datasets, which potentially produce different conservation results when developing conservation plans for adaptation to SLR. The research results show that a large number of additional lands can be identified as being subject to SLR when vertical accuracy of elevation data is considered. Species habitat and conservation prioritization for adaptation to SLR could be significantly different with different coastal impact models and elevation dataset combinations. The research suggests that it is valuable to invest in more sophisticated coastal impact models and high-resolution elevation datasets because the less accurate model and elevation dataset could fail to identify areas of high conservation value in the face of SLR.

It has been recognized that we should use better data and coastal impact models to assess SLR impacts (Poulter & Halpin, 2007; Gesch, 2009). Nevertheless, this does not mean that we should use the more accurate model and dataset under all situations. There is an increasing availability of Lidar elevation data in coastal areas of the United States, and there is an interest in developing a national collection of Lidar elevation data (Stoker et al., 2007; Gesch, 2009). However, we should realize that in some data-poor areas of the world this high quality elevation data is not always obtainable. In addition, running the more sophisticated coastal impact models such as SLAMM is costly and requires expert knowledge which is not available when the project budget is limited. One of the advantages of bathtub models in assessing SLR impacts is that it could give a sense of the magnitude of areas that would be impacted by SLR. Choosing which model is to be used in SLR impact assessment depends on the goal of the planning or research. I would recommend using the most sophisticated model and high-resolution elevation dataset in the identification of conservation priorities in response to SLR while using the bathtub model for SLR vulnerability assessment. Bathtub based assessments, including ones using coarser DEM data, can still be useful when better data and models are not available or feasible, but results need to be appropriately assessed and communicated.

Table 4-1 Model-dataset combinations

Model – dataset Combination	Coastal Impact Model	Elevation Dataset	Description
SLAMM -Lidar	SLAMM	Lidar (10m)	The best available DEM data with process-based model can predict coastal habitat change in response to SLR based on key ecological processes and abiotic factors.
Bathtub-Lidar	Bathtub	LIDAR (10 m)	The best available DEM data with simple bathtub inundation analysis. Sites below the given elevation will be inundated.
Bathtub-NED	Bathtub	NED (30m)	The coarse input data with simple bathtub inundation analysis. This is the cheapest and fastest option but omits key ecological processes such as hydrological process.

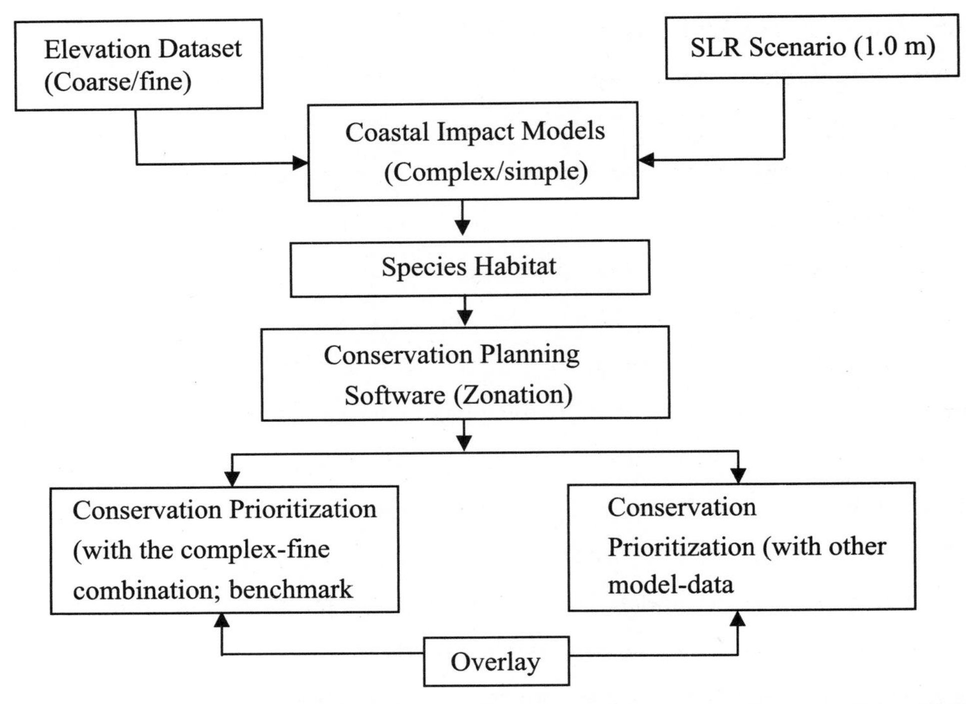

Figure 4-1 Diagram of the methodology used to compare conservation outcomes with different model-dataset combinations

Table 4-2 Characteristics of elevation datasets analyzed in the study area

Elevation Data	Resolution	RMSE	Linear error at 95% CI
NED	30 m	1.27 m	±2.21 m
Lidar DEM	10 m	0.14 m	±0.27 m

Table 4-3 Statistics of potential inundation areas to a 1.0 m SLR with / without consideration of elevation uncertainty

Elevation data	Inundation area without consideration of elevation uncertainty (acres)	Inundation area with consideration of elevation uncertainties (acres)	Increase(%)
NED	38,436.78	58,707.18	52.74%
Lidar	27,022.53	29,627.77	9.64%

Table 4-4 Estimated areas of wetlands and water distribution under different model-dataset combinations

Current	Bathtub-NED	Bathtub-Lidar	SLAMM-Lidar	
Wetlands (Acres)	89,747.21	71,779.61	73,024.75	92,056.44
Water (Acres)	12,703.86	41,836.78	30,256.85	20,149.28

Table 4-5 Statistics of species habitat under different model-dataset combinations

	SpeciesCurrent	Bathtub-NED (Acres)	Bathtub-Lidar (Acres)	SLAMM-Lidar (Acres)
Black rail	1,601.00	938.51	943.18	679.50
Limpkin	6,003.00	4,863.89	5178.28	5,264.44
Painted bunting	9,463.00	7,144.28	7,952.15	8,962.44
Oystercatcher	3,444.40	0	0	9207.88
Marsh Wren	6,792.41	0	0	5,551.20

CHAPTER 5
SUMMARY, POLICY ISSUES, AND RECOMMENDATIONS

5.1 Summary of the Research and Findings

The three primary questions addressed in this research comprise some of the most urgent research needs for adaptation to SLR, with particular relevance to Florida. Firstly, chapter two addresses the question of "where existing conservation areas should be enlarged, where corridors should be established and where new conservation areas should be created" (Noss, 2011, p10) to facilitate biodiversity adaptation to SLR. Secondly, chapter three addresses the question of how economic concepts could be incorporated into SLR adaptive conservation planning and reserve design to achieve conservation efficiency. Thirdly, chapter four addresses the question of what is the value of information (model sophistication and data accuracy) for SLR adaptive conservation planning and reserve design.

5.1.1 Sea Level Rise Adaptive Conservation Planning and Reserve Design

SLR is an important consideration to be included in coastal biodiversity conservation, yet many conservation planning and reserve design approaches have not yet taken SLR into account (Runting et al., 2012; Wetzel et al., 2012). There are several major reasons for this. Firstly, the uncertainties associated with SLR make the prediction of SLR impacts inherently difficult. Secondly, there is a lack of improved models, predictions and understanding of SLR impacts on species, natural communities and ecosystems (Noss, 2011). Though there are difficulties, incorporating SLR into coastal conservation planning and reserve design is critical to protect biodiversity under landscape dynamics due to SLR and land use change.

The integrated SLR adaptive conservation planning process that incorporates a geomorphological model (SLAMM), species habitat model, and conservation prioritization software Zonation developed in this dissertation provides one method that takes SLR into account when designing reserve networks. Application of this integrated model in the Matanzas river basin of Northeast Florida indicates that current conservation and managed areas are not sufficient to protect biodiversity as sea level rises and land use changes. Functional adaptation will require that existing conservation areas be expanded to respond to these landscape dynamics.

My research results also suggest that existing conservation planning tools in Florida serve as a good foundation for directing additional conservation efforts to facilitate adaptation to SLR. Comparison of the conservation priorities identified by Zonation with the most recent version of FEGN indicates that FEGN could serve as a good guide for identifying adaptive strategies in response to SLR, especially in regards to protecting intact coastal to inland corridors. Florida has been ahead of many other states and regions in addressing the protection of ecological connectivity, and this work provides an important foundation for SLR conservation planning. Comparison of the conservation priorities identified by Zonation with the most recent version of the CLIP data indicates that CLIP can also be used at least as a starting point for identifying conservation priorities relevant to SLR adaptation, and it suggests that, lacking other data, the FEGN and CLIP data could serve as a surrogate for more detailed assessments of conservation priorities relevant to SLR adaptation. In addition, top conservation priorities in response to SLR identified in this research could be used to augment existing conservation layers such as FEGN and CLIP to fully incorporate SLR considerations.

5.1.2 Economic Concepts in Conservation Planning and Reserve Design

Most studies of conservation planning and reserve design have focused on the biological benefits of conservation. However, the economic costs of conservation should not be ignored (Naidoo et al., 2006). Understanding the economic benefits and costs of conservation will help conservation decision makers to allocate limited conservation funds and resources in a more efficient manner (Naidoo & Ricketts, 2006). The need for conservation efficiency provides conservation planners an important rationale to incorporate economic concepts into SLR adaptive conservation planning and reserve design.

The dissertation presents a case study to demonstrate how economic concepts (ecosystem services valuation and cost of conservation) can be incorporated into SLR adaptive conservation planning and reserve design. The analysis showed that explicit analysis of economic costs and benefits of conservation could inform conservation decision-making. Coastal reserve managers will need this information to allocate their funds and resources more strategically and efficiently to achieve the conservation goal of facilitating biodiversity adaptation to future SLR. However, costs still need to be incorporated carefully in the planning process because conservation priorities may be important to protect regardless of relative cost.

5.1.3 Impacts of Model and Elevation Data Accuracy in Sea Level Rise Adaptive Conservation Planning and Reserve Design

SLR adaptive conservation planning and reserve design relies on the sophistication of coastal impact models and the accuracy of elevation data to identify conservation priorities in response to SLR. Therefore, conservation planners and decision makers need to know the impacts of model and elevation data accuracy in adaptive conservation designs in response to SLR. This dissertation compared three different model-dataset combinations (Bathtub-NED, Bathtub-Lidar and SLAMM-Lidar) in identifying SLR adaptive conservation priorities in the Matanzas study area and found that using different coastal impact models and elevation datasets could clearly lead to significantly different conservation prioritization outcomes. Using the less accurate model and elevation datasets in the identification of SLR adaptive conservation priorities could fail to identify areas of high conservation value for adaptation to SLR. This suggests that it is valuable to invest in more sophisticated coastal impact model and high-resolution elevation datasets when developing conservation plans for biodiversity conservation in the face of landscape dynamics due to SLR. However, it could still be appropriate to use less accurate coastal impact models and elevation datasets when better data and models are not available or feasible. The less resolute models and data could give a sense of the magnitude of areas that would be impacted by SLR.

5.2 Shortcomings and Future Research

Analysis in this dissertation provides answers for these research questions, yet the analysis has shortcomings that future research could address. The identification of conservation priorities in response to SLR and land use change depends on potential species habitat derived from species habitat models and only those 38 vertebrate focal species were included in the analysis. Future research could include aquatic species in the list when adequate species habitat data or models are available. In addition, species habitat only represents one aspect of biodiversity that could be used to identify SLR adaptive conservation priorities. Other biodiversity features such as natural communities were not incorporated in this analysis. However, natural communities could also be used as an appropriate biodiversity feature to identify conservation priorities (Arponen et al., 2008; Leathwick et al., 2010), especially in regards to addressing the "coarse filter" part of reserve design recommendations (Hunter, 1990). I have focused on various vertebrate focal species habitat models, but addition of natural communities could address other species including many plants and invertebrates that might not be covered by the focal species selected. Similar to species habitat, natural communities such as upland hardwood forest and tidal flat will also change in response to the rising sea water.Future research could identify conservation priorities in response to SLR at the natural community level or a combined community and species level.

SLAMM was used to simulate marsh alteration and migration due to SLR and it was used as a foundation for running species habitat models and then identifying conservation priorities in response to SLR. However, other changes such as potential changes in upland habitats from dry to more mesic as sea level rises is not addressed by SLAMM. In addition, SLAMM only takes SLR impacts on coastal wetlands into account; other factors that could also impact coastal habitat change such as temperature increases or precipitation changes are not accounted for in the modeling process. For example, SLAMM projects that mangrove will decrease significantly as sea level rises. However, mangrove could also increase because of potential temperature increase as a result of global warming, and replace a lot of future saltmarsh habitat, which means that impacts on saltmarsh dependent species might be significantly under-represented in these analyses. For a more comprehensive understanding of climate change impacts on coastal habitats, more detailed models of ecosystem structure and function changes would be useful, including the possibility of coupling SLAMM modeling with projected changes in vegetation based on downscaled climate models.

The incorporation of economic concepts into SLR adaptive conservation planning considered conservation costs and benefits. Cost of conservation includes land acquisition costs, management costs, transaction costs,damage costs and opportunity costs (Naidoo et al., 2006). Due to data limitations, this research only included land acquisition costs in the analysis. For a more comprehensive estimation of conservation costs, it would be useful to include all five costs in the analysis. Benefits of conservation were estimated based on the Gecoserv database that is relevant to the Gulf of Mexico region. The estimation of ecosystem services valuation in the Matanzas study area was based on this database but we should understand that the ecosystem services value in the Atlantic coast could be different from that in the Gulf of Mexico coast. The average value of each service provided by coastal habitats was estimated based on the published research conducted by researchers using different models. The reliability of ecosystem services valuation in this research totally depends on the accuracy of the economic models used by the researchers. In addition, only ecosystem services that could be found in the database were counted in this research and ecosystem services of upland were only very coarsely incorporated in the analysis. This might underestimate the services provided by coastal habitats and potentially lost to SLR. For a more comprehensive understanding of the changing ecosystem services provided by coastal habitats, other services should also be included when they are available in the database. For example, The InVEST software developed by the Natural Capital Project (http://www.naturalcapitalproject.org/InVEST.html) could be used to assess the integrated valuation of environmental services provided by coastal habitats including coastal protection, water quality, carbon sequestration, recreation, etc. This software enables decision makers to assess quantified tradeoffs associated with alternative management choices and to identify areas where investment in natural capital can enhance human development and conservation (Tallis et al., 2013). With more funding and time, using this tool in the research could significantly increase the detail of ecosystem services impacts.

The comparison of conservation outcomes using different model-dataset combinations in reserve design demonstrated the value of more detailed information (model sophistication and data accuracy) in SLR adaptive conservation planning and reserve design. Because of timing and funding, this research only compared conservation outcomes with three different model-dataset combinations including Bathtub-NED, Bathtub-Lidar and SLAMM-Lidar but the SLAMM-NED combination was not included in the analysis. Including the SLAMM-NED combination in the analysis would add to the existing research and present the conservation outcomes with a complex coastal impact model and coarse elevation data. Compared to conservation outcomes with the SLAMM-Lidar combination, this added analysis would likely demonstrate the value of more detailed elevation data in identifying SLR adaptive conservation priorities while using the complex coastal impact model.

Models and software used in the dissertation research include SLAMM, species habitat models and Zonation. These models are all well designed and evaluated by the model developers to ensure their accuracy and usefulness. Nevertheless, we should understand that "essentially, all models are wrong, but some are useful" (Box &Draper, 1987). Models used in this research are useful but we should recognize that these models have limitations and can never precisely simulate the actual situations under SLR, thus the conservation priorities identified by the integrated modeling process will never be the perfect solution. As part of future improvements in SLR adaptive conservation planning and reserve design, the conservation prioritization identified in this research should be periodically updated to reflect new information as model sophistication and data accuracy improves.

5.3 Land Conservation Policy Issues for Adaptation to Sea Level Rise

5.3.1 Land Protection

Florida has been a leader in the pursuit of biodiversity protection. Florida has more than 12 million acres of lands protected and managed predominantly or partly for conservation, which account for more than 20% of the total state lands and water (FDEP, 2011). Preservation 2000 and Florida Forever, the state's premier conservation lands acquisition programs that began back in 1990, have protected more than 2.5 million acres for conservation (FDEP, 2011). Since its inception in July 2001, the Florida Forever program has spent approximately $3 billion on conservation land acquisition in Florida. Unfortunately, since the economic downturn in 2008, the Governor and Legislature have significantly reduced the funding for Florida Forever. The funding cut could do significant harm to biodiversity conservation efforts including SLR mitigation and adaptation. The Florida Water and Land Conservation Initiative Amendment 1which aims to delicate funds for conservation lands acquisition could be used to fully fund Florida Forever again, and if so, SLR adaptive conservation opportunities should be a high priority.

Coupled with rapid land use change, SLR will likely constitute one of the most fundamental and important challenges for biodiversity conservation in Florida (Hoctor et al., 2014). Conservation lands acquisition programs such as Florida Forever will need to meet this new challenge to conserve biodiversity under a changing climate. Figure 2-11 shows that additional lands will need to be acquired to facilitate biodiversity adaptation to SLR. To acquire these lands, "funding should be restored to critical land acquisition programs such as the Florida Forever" (Noss, 2011, p13). To prevent further loss of biodiversity to SLR and land use change, "conservation land acquisition (including easements) will need to be greatly accelerated to expand boundaries of coastal conservation areas upslope and inland, establish broad coast-inland habitat corridors, and protect inland refugia" (Noss, 2011, p13).

Other federal and state conservation programs and incentives other than Florida Forever that could be relevant to biodiversity conservation under SLR in Florida include conservation incentive programs in FWC's Cooperative Conservation Blueprint (CCB) project which aims to create a multi-partner strategic conservation process to implement the Florida Wildlife Action Plan (FWC, 2014), the Rural and Family Lands Protection Act (Florida Department of Agriculture and Consumer Services, 2013), the Wetland Reserve Program (WRP) from Natural Resources Conservation Services (United States Department of Agriculture, 2012), the Landscape Conservation Cooperatives (LCCs) from U.S. Fish and Wildlife Services (USFWS, 2014), etc. In particular, the Peninsular Florida Landscape Conservation Cooperative (PFLCC), which aims to support conservation in Florida at the landscape scale, lists SLR as one of its most critical challenges to be addressed (FWC, 2010). Ensuring funding and support for these conservation programs is essential for the protection of SLR adaptation opportunities in the state (Noss et al., 2014).

Interagency coordination is crucial in conservation programs and incentives. To gain and maintain support from state agencies and non-governmental organizations, the CCB is looking opportunities for partnerships such as integrating with the PFLCC initiative to seek funding and resources. Expanding existing partnerships among stakeholders including federal, state and local agencies, non-profit organizations, universities, private landowners and others is one of the major goals of PFLCC (USFWS, 2010). As PFLCC addresses SLR in its efforts, ensuring funding and multi-partner support of PFLCC would greatly benefit biodiversity SLR adaptation in Florida.

5.3.2 Growth Management

Growth management is a set of techniques that is used by governments to guide development in order to promote sustainable and vibrant communities. In 1985, Florida enacted the Growth Management Act that required local governments to create comprehensive plans to guide development. However, the legal requirements have largely failed (Nicolas & Steiner, 2000), and the requirements have been weakened. This is because current local comprehensive plans appear to accommodate growth and do not protect critical natural resources (Hoctor, 2003). As sea level rises, growth management in Florida will need to meet some new challenges. Firstly , areas that are not designated for conservation might in fact be critical habitat or migration corridors for some species as sea level rises thus growth management will need to guide development to avoid these areas. Secondly, mass human migration from the coasts to inland areas will likely happen after some SLR threshold is reached (Noss, 2011). This migration might conflict with species or species habitat migration as sea level rises. To address this population shift, growth management in Florida will need to be re-examined to reflect this change. In recent years, Florida's growth management has undergone significant changes that reduced the state's oversight of local governments (Florida Association of Counties, 2013). This change of growth management in Florida could significantly limit the ability of planning to avoid impacts to conservation priorities.

The comprehensive land use plan is an important planning tool to regulate land use in an efficient and fair way. Florida's Growth Management Act requires local counties to adopt their comprehensive land use plans to guide the orderly and balanced future development of their areas (FS Ch. 163.3177). Though the current state law does not require local governments to consider SLR in comprehensive plans, land use planning and SLR should be intricately linked in Florida.

To help mitigate SLR impacts, the Florida legislature will need to consider amending state planning statutes to include more explicit consideration of SLR. "Florida Resilient Coasts", a Florida Atlantic University (FAU) report indicates that "Florida's comprehensive planning laws and procedures will need careful re-examination in light of climate change predictions……" (FAU, 2008). Particularly, the state policies should be amended to ensure that future development will not put coastal communities that are likely to experience SLR impacts in harm's way (FAU, 2008). State law requiring consideration of SLR will place heightened responsibility on local governments to include SLR considerations in their comprehensive plans.

Future land use plans should be revised to incorporate SLR assessments (FAU, 2008). "Understanding of how land use planning (e.g., county comprehensive plans) should be revised to include explicit consideration of SLR" is considered one of the most important research needs for adaptation to SLR in Florida (Noss, 2011, p10). SLR adaptive conservation priorities could be incorporated into local comprehensive plans to facilitate biodiversity adaptation to future environmental changes. For instance, the top conservation priorities identified in this research could be incorporated into local county (St. Johns and Flagler counties) comprehensive plans to assist SLR adaptation at the local level. As local counties need to review their comprehensive land use plans to determine whether it is necessary to update comprehensive plans to meet new challenges, this provides a natural opportunity for local governments to include SLR considerations in the plans to protect those areas of high biodiversity conservation value for adaptation to SLR.

Green infrastructure approaches could assist coastal communities to achieve the goal of urban sustainability and coastal resilience (Foster et al., 2011). The green infrastructure based planning approach is suggested by the U.S. Environmental Protection Agency (EPA) to build coastal resilience in response to SLR (EPA, 2014). Compared to grey infrastructure approaches that emphasize building hard structures such as sea walls and bulkheads, green infrastructure approaches use natural systems such as native wetlands and coastal dunes to deal with SLR impacts. A living shoreline which can be a mixture of saltwater marshes, oyster reefs, tidal flats, etc., is a good example of a green infrastructure approach for SLR mitigation and adaptation. As sea level rises, conservation priorities identified for adaptation to SLR are "green infrastructure" that needs to be protected to avoid future development.

High Velocity Hurricane Zones (HVHZ) are relevant to state policies about SLR adaptation. The HVHZ was established to defend against the impacts from high speed wind and it basically ignores the impacts from water which is much denser than air at sea level (Alvarez, 2012). As sea level rises, the impacts from sea water in HVHZ will increase. The current state government considers HVHZ to be of critical state concern warranting special attention when counties propose comprehensive plan changes. Including considerations of SLR impacts in HVHZ and incorporating HVHZ in local comprehensive plans to discourage new development with HVHZ will assist SLR adaptation because these zones might be of critical importance for biodiversity adaptation to SLR.

Florida will need to adopt a smarter growth pattern that avoids, minimizes and mitigates the impacts of SLR. Planning future development in a careful and smart growth oriented manner to avoid competing with SLR adaptive conservation priorities is a key component for land use decision makers to consider. The future impact of additional development could be avoided or minimized with efforts to change growth patterns including increasing residential densities. Adopting proactive protection measures including changing development patterns and densities could contribute significantly to protection of the top SLR adaptive conservation priorities to facilitate biodiversity adaptation in response to future SLR.

5.4 Transferability to Other Coastal Regions and Integration to Other Coastal Decision Tools

The integrated modeling and conservation prioritization approach is useful, robust, and transferable to other Florida coastal regions and other low-lying coastal areas of the world to plan for SLR adaptation. To ensure transferability, the first step is to conduct a more detailed regional analysis of other high priority SLR impacts areas such as southwest Florida and the Big Bend coast that need to be protected to effectively conserve biodiversity in the face of landscape dynamics due to SLR and land use change in Florida (Hoctor et al., 2014; Noss et al., 2014). The next step is to scale up the research to the statewide level to facilitate statewide efforts for biodiversity adaptation to SLR. The statewide SLR adaptive conservation priorities could be used by state agencies and local governments to guide conservation lands acquisition for adaptation to SLR.

To maximize impacts of this research, seeking opportunities to integrate SLR adaptive conservation priorities into other coastal decision support tools is important. NOAA's Sea Level Rise and Coastal Flooding Viewer (http://www.csc.noaa.gov/slr/viewer/) is a teaching and planning tool that visualizes potential impacts from SLR including potential marsh migration due to the rising sea water. Similar to the existing categories including SLR, vulnerability, and flood frequency that could be visualized in this viewer, conservation priorities adaptive to SLR (outputs of Zonation) are potentially feasible to be included in this tool to allow coastal decision makers to visualize the change of conservation priorities under SLR scenarios. In addition, The Nature Conservancy's Gulf of Mexico Coastal Resilience Tool (http://maps.coastalresilience.org/gulfmex/#) is another potential opportunity that SLR adaptive conservation prioritization could be integrated. The Gulf of Mexico Coastal Resilience Tool is an interactive tool that helps inform conservation and restoration efforts throughout the Gulf of Mexico. Similar to future habitat under different SLR scenarios that could be visualized via this tool, conservation priorities that are adaptive to SLR could also be added to this tool as a layer. This integration will provide the Gulf region spatially-explicit information about where to conserve in order to achieve the conservation goal of facilitating biodiversity adaptation to future SLR.

APPENDIX A
SPECIES HABITAT MODELING DESCRIPTION FOR AMERICAN OYSTERCATCHER

Summary

The American Oystercatcher (*Haematopus palliates*) is a threatened species that can be found on both the Gulf and Atlantic coasts of Florida. Habitats for feeding and roosting include shell bars, spoil islands, coastal beaches, mudflats and salt marshes. The American Oystercatcher prefers large sand areas with sparse vegetation for nesting.

Model Steps

1. Appropriate American Oystercatcher foraging and roosting habitat was identified as Non-Vegetated Wetlands, Tidal Flats, Shorelines, Oyster bars, Beaches Other Than Swimming Beaches, and Exposed Rock with Marsh Grasses in managed areas where it has been found to occur.

2. Spoil areas adjacent to estuarine waters, saltmarsh, or mangrove were included.

3. Also identified all saltwater within 30 meters of saltmarsh and mangroves and within 1 kilometer of habitat identified in steps 1 and 2.

4. The final habitat map is created by giving all potential habitat identified in steps 1-3 a value of 1 and all other cells a value of 0.

Model Script

1. Find only appropriate land uses within 1 km from estuarine waters. This eliminates the potential for inappropriate or misclassified land uses away from estuarine habitat to be

included in the model.

"[out1] = CON([landuse] == 5400 or [landuse] == 5410 or [landuse] == 5420 or [landuse] == 5430 or [landuse] == 5700, 1)" + vbLf + _

"[out2] = EUCDISTANCE([out1], #, #, 1000)" + vbLf + _

"[out3] = CON(isnull([out2]) and [landuse] ge 0, 0, CON([out2] le 1000, 1, 0))" + vbLf + _

2. Find appropriate American Oystercatcher foraging and nesting habitat: 6500 – Non-Vegetated Wetlands, 6510 – Tidal Flats, 6520 – Shorelines, 6540 – Oysterbars, 7100 – 7110 – Beaches Other Than Swimming Beaches, 7200 – Other Sand, 7310 – Exposed Rock With Marsh Grasses.

"[out4] = CON([out3] == 1 and ([landuse] == 6500 or [landuse] == 6510 or [landuse] == 6520 or [landuse] == 6540 or [landuse] == 7100 or [landuse] == 7110 or [landuse] == 7200 or [landuse] == 7310), 1, 0)" + vbLf + _

3. Spoil Areas (7430) and Disturbed Land (7400) adjacent to estuarine waters and open salt water and saltmarsh or mangrove.

"[out5] = CON([landuse] == 5400 or [landuse] == 5410 or [landuse] == 5420 or [landuse] == 5430 or [landuse] == 5700 or [landuse] == 6120 or ([landuse] ge 6420 and [landuse] lt 6430), 1, 0)" + vbLf + _

"[out6] = CON([landuse] == 7430 or [landuse] == 7400, 1)" + vbLf + _

"[out7] = EXPAND([out5], 1, list, 1)" + vbLf + _

"[out8] = REGIONGROUP([out6], #, eight, within)" + vbLf + _

"[out9] = ZONALMAX([out8], [out7])" + vbLf + _

"[out10] = CON(isnull([out9]) and [landuse] ge 0, 0, CON([out9] == 1, 1, 0))" + vbLf + _

4. Add appropriate spoil areas to OUT4.

"[out11] = CON([out4] == 1 or [out10] == 1, 1, 0)" + vbLf + _

5. Add estuarine/saltwater bodies within 30 meters of salt marshes and mangrove edges and under 1 kilometer from primary foraging, nesting, roosting habitat (step 2).

"[out12] = EXPAND(CON([landuse] == 5400 or [landuse] == 5410 or [landuse] == 5420 or [landuse] == 5430 or [landuse] == 5700, 1, 0), 1, list, 1)" + vbLf + _

"[out13] = EXPAND(CON([landuse] == 6120 or ([landuse] ge 6420 and [landuse] lt 6430), 1, 0), 1, list, 1)" + vbLf + _

"[out14] = con((([landuse] == 5400 or [landuse] == 5410 or [landuse] == 5420 or [landuse] == 5430 or [landuse] == 5700) and [out13] == 1) or (([landuse] == 6120 or ([landuse] ge 6420 and [landuse] lt 6430)) and [out12] == 1), 1)" + vbLf + _

"[out15] = EUCDISTANCE([out14], #, #, 30)" + vbLf + _

"[out16] = CON(isnull([out15]) and [landuse] ge 0, 0, CON([out15] le 30 and ([landuse] == 5400 or [landuse] == 5410 or [landuse] == 5420 or [landuse] == 5430 or [landuse] == 5700), 1, 0))" + vbLf + _

"[out17] = EUCDISTANCE(CON([out11] == 1, 1), #, #, 1000)" + vbLf + _

"[out18] = CON(isnull([out17]) and [landuse] ge 0, 0, CON([out17] le 1000 and [out16] == 1, 1, 0))" + vbLf + _

6. Combine step 5 ([out17]) and step 4 ([out11]) to give a final American Oystercatcher habitat model.

"[out19] = CON([out11] == 1 or [out18] == 1, 1, 0)"

"[out1] = CON([landuse] == 5400 or [landuse] == 5410 or [landuse] == 5420 or [landuse] == 5430 or [landuse] == 5700, 1)" + vbLf + _

"[out2] = EUCDISTANCE([out1], #, #, 1000)" + vbLf + _

"[out3] = CON(isnull([out2]) and [landuse] ge 0, 0, CON([out2] le 1000, 1, 0))" + vbLf + _

"[out4] = CON([out3] == 1 and ([landuse] == 6500 or [landuse] == 6510 or [landuse] == 6520 or [landuse] == 6540 or [landuse] == 7100 or [landuse] == 7110 or [landuse] == 7200 or [landuse] == 7310), 1, 0)" + vbLf + _

"[out5] = CON([landuse] == 5400 or [landuse] == 5410 or [landuse] == 5420 or [landuse] == 5430 or [landuse] == 5700 or [landuse] == 6120 or ([landuse] ge 6420 and [landuse] lt 6430), 1, 0)" + vbLf + _

"[out6] = CON([landuse] == 7430 or [landuse] == 7400, 1)" + vbLf + _

"[out7] = EXPAND([out5], 1, list, 1)" + vbLf + _

"[out8] = REGIONGROUP([out6], #, eight, within)" + vbLf + _

"[out9] = ZONALMAX([out8], [out7])" + vbLf + _

"[out10] = CON(isnull([out9]) and [landuse] ge 0, 0, CON([out9] == 1, 1, 0))" + vbLf + _
"[out11] = CON([out4] == 1 or [out10] == 1, 1, 0)" + vbLf + _
"[out12] = EXPAND(CON([landuse] == 5400 or [landuse] == 5410 or [landuse] == 5420 or [landuse] == 5430 or [landuse] == 5700, 1, 0), 1, list, 1)" + vbLf + _
"[out13] = EXPAND(CON([landuse] == 6120 or ([landuse] ge 6420 and [landuse] lt 6430), 1, 0), 1, list, 1)" + vbLf + _
"[out14] = con((([landuse] == 5400 or [landuse] == 5410 or [landuse] == 5420 or [landuse] == 5430 or [landuse] == 5700) and [out13] == 1) or (([landuse] == 6120 or ([landuse] ge 6420 and [landuse] lt 6430)) and [out12] == 1), 1)" + vbLf + _
"[out15] = EUCDISTANCE([out14], #, #, 30)" + vbLf + _
"[out16] = CON(isnull([out15]) and [landuse] ge 0, 0, CON([out15] le 30 and ([landuse] == 5400 or [landuse] == 5410 or [landuse] == 5420 or [landuse] == 5430 or [landuse] == 5700), 1, 0))" + vbLf + _
"[out17] = EUCDISTANCE(CON([out11] == 1, 1), #, #, 1000)" + vbLf + _
"[out18] = CON(isnull([out17]) and [landuse] ge 0, 0, CON([out17] le 1000 and [out16] == 1, 1, 0))" + vbLf + _
"[out19] = CON([out11] == 1 or [out18] == 1, 1, 0)"

APPENDIX B
FOCAL SPECIES IN THE MATANZAS RIVER BASIN STUDY AREA

Table B-1 Focal species in the matanzas river basin study area

Common Name	Scientific Name	Global Rank	State Rank	Federal Status	State Status
Gopher Tortoise	Gopherus polyphemus	G3	S3	C	ST
Spotted Turtle	Clemmys guttata	G5	S3	N	N
Eastern Indigo Snake	Drymarchon couper	G3	S3	LT	FT
Diamondback Rattlesnake	Crotalus adamanteus	G4	S3	N	N
Florida Kingsnake	Lampropeltis betula floridan				
Florida Pine Snake	Pituophis melanoleucus mugitus	G4T3	S3	N	SSC
Sandhill Crane	Grus canadensis pratensis	G5T2T3	S2S3	N	ST
Gopher Frog	Rana capita	G3	S3	N	SSC
Southeastern American Kestrel	Falco sparverius paulus	G5T4	S3	N	ST
Neotropical Migrant Forest Bird Guild					
Migratory (Wintering) Waterfowl					
Wading Bird Guild					
Black Rail	Laterallus jamaicensis	G4	S2	N	N
Wood Stork	Mycteria americana	G4	S2	LE	FE
Swallow-tailed Kite	Elanoides forficatus	G5	S2	N	N

Continued

Common Name	Scientific Name	Global Rank	State Rank	Federal Status	State Status
Bald Eagle	Haliaeetus leucocephalus	G5	S3	N	N
Limpkin	Aramus guarauna	G5	S3	N	SSC
Striped Newt	Notophthalmus perstriatus	G2G3	S2S3	C	N
American Oystercatcher	Haematopus palliatus	G5	S2	N	SSC
Florida Scrub-Jay	Aphelocoma coerulescens	G2	S2	LT	FT
Bachman's Sparrow	Peucaea aestivalis	G3	S3	N	N
Round-tailed Muskrat	Neofiber alleni	G3	S3	N	N
Florida Mink	Neovison bison	G5T3	S3	N	N
River Otter	Lontra canadensis				
Florida Mouse	Podomys floridanus	G3	S3	N	SSC
Sherman's Fox Squirrel	Sciurus niger shermani	G5T3	S3	N	SSC
Florida Black Bear	Ursus americanus floridanus	G5T2	S2	N	ST*
Sea Turtles					
Shorebird Guild- Open water foraging					
Shorebird Guild- Sand foraging					
Painted Bunting	Passerina ciris	G5	S3	N	N
Anastasia Beach Mouse	Peromyscus polionotus phasma	G5T1	S1	LE	FE
Merlin					
Gulf Saltmarsh Snake (for Atlantic Saltmarsh Snake)					
Mangrove Forest Bird Guild (Florida Prairie Warbler)					

Continued

Common Name	Scientific Name	Global Rank	State Rank	Federal Status	State Status
Marian's Marsh Wren (for Worthington's Marsh Wren)					
Ornate Diamondback Terrapin (for Diamondback Terrapin)					
Scott's Seaside Sparrow (for MacGillivray's seaside sparrow)					

Table B-2 The avalue for focal species in the Matanzas river basin study area

Common Name	Relative Weight (4 is the highest, 1 is the lowest)	Dispersal Capability (m)	α Value in Zonation (α=2/Dispersal Capability)
Gopher Tortoise	2	1,000	0.002
Spotted Turtle	2	2,000	0.0010
Eastern Indigo Snake	2	2,000	0.001
Diamondback Rattlesnake	2	1,000	0.002
Florida Kingsnake	1	5,000	0.0004
Florida Pine Snake	2	500	0.004
Sandhill Crane	3	10,000	0.0002
Gopher Frog	2	1,000	0.002
Southeastern American Kestrel	2	8,000	0.00025
Neotropical Migrant Forest Bird Guild	3	5,000	0.0004
Migratory (Wintering) Waterfowl	1	10,000	0.0002
Wading bird Guild	3	15,000	0.000133

Continued

Common Name	Relative Weight (4 is the highest, 1 is the lowest)	Dispersal Capability (m)	α Value in Zonation (α=2/Dispersal Capability)
Black Rail	3	280	0.007143
Wood Stork	3	15,000	0.000133
Swallow-tailed Kite	3	10,000	0.00020
Bald Eagle	2	10,000	0.0002
Limpkin	2	5,000	0.0004
Striped Newt	3	1,000	0.002
American Oystercatcher	3	5,000	0.0004
Florida Scrub-Jay	3	3,500	0.00057
Bachman's Sparrow	2	150	0.01333
Round-tailed Muskrat	2	1,000	0.00200
Florida Mink	2	4,000	0.0005
River Otter	1	36,000	5.55556E-05
Florida Mouse	2	2,000	0.001
Sherman's Fox Squirrel	2	1,000	0.002
Florida Black Bear	3	30,000	6.66667E-05
Sea Turtles	4	4,000	0.0005
Shorebird Guild- Open water foraging	2	5,000	0.0004
Shorebird Guild- Sand foraging	4	1,500	0.00133
Painted Bunting	2	5,000	0.0004
Anastasia Beach Mouse	4	40	0.05

Continued

Common Name	Relative Weight (4 is the highest, 1 is the lowest)	Dispersal Capability (m)	α Value in Zonation (α=2/Dispersal Capability)
Merlin	3	10,000	0.0002
Atlantic Saltmarsh Snake	4	1,000	0.002
Florida Prairie Warbler	2	250	0.008
Worthington's Marsh Wren	2	150	0.013
Diamondback Terrapin	1	1,500	0.00133
MacGillivray's seaside sparrow	2	1,200	0.002

APPENDIX C
THE RUN SETTINGS FILE USED IN ZONATION

```
[Settings]
removal rule = 1
warp factor = 100
edge removal = 1
add edge points = 0
use SSI = 0
SSI file name = tutorial_input/SSI_list.txt
use planning unit layer = 0
planning unit layer file = tutorial_input/plu.asc

use cost = 0
cost file = tutorial_input/cost.asc
use mask = 1
mask file = tutorial_input/mask_towns.tif
use boundary quality penalty = 0
BQP profiles file = tutorial_input/BQPcurves.txt
BQP mode = 1
BLP = 0
use tree connectivity = 0
tree connectivity file = tutorial_input/tree.txt
use interactions = 0
interaction file = tutorial_input/interact.spp

annotate name = 0
logit space = 0
treat zero-areas as missing data = 0
```

z = 0.25

resample species = 0

[Info-gap settings]

Info-gap proportional = 0

use info-gap weights = 0

Info-gap weights file = tutorial_input/UCweights.spp

APPENDIX D

SPECIES HABITAT CHANGE DUE TO SEA LEVEL RISE AND LAND USE CHANGE

Common Name	Current Habitat(Acres)	Net Gain/Loss to 0.5 m SLR	Percent Gain/ Loss to 0.5 m SLR	Loss to Land Use Change (Acres)	Percent Loss to Land Use Change	Total Percent Gain/Loss to 0.5 m SLR and Land Use Change
Gopher Tortoise	13,346	-981	-7.35%	-2,642	-19.79%	-27.14%
Spotted Turtle	104,397	-321	-0.31%	-18,749	-17.96%	-18.27%
Eastern Indigo Snake	180,685	4,558	2.52%	-34,184	-18.92%	-16.40%
Diamondback Rattlesnake	139,379	-144	-0.10%	-28,038	-20.12%	-20.22%
Florida Kingsnake	42,407	976	2.30%	-28,038	-66.12%	-63.82%
Florida Pine Snake	12,587	-988	-7.85%	-2530.45	-20.10%	-27.96%
Sandhill Crane	41,369	-1,220	-2.95%	-8279.00	-20.01%	-22.96%
Gopher Frog	68,501	-16,402	-23.94%	-10961.28	-16.00%	-39.95%
Southeastern American Kestrel	15,118	-1,927	-12.75%	-3704.55	-24.50%	-37.25%
Neotropical Migrant						
Forest Bird Guild	36,180	-1,029	-2.84%	-5279.24	-14.59%	-17.43%
Migratory (Wintering) Waterfowl	11,688	22	0.19%	-163.48	-1.40%	-1.21%

Continued

Common Name	Current Habitat(Acres)	Net Gain/Loss to 0.5 m SLR	Percent Gain/ Loss to 0.5 m SLR	Loss to Land Use Change (Acres)	Percent Loss to Land Use Change	Total Percent Gain/Loss to 0.5 m SLR and Land Use Change
Wading bird Guild	94,427	6,176	6.54%	−12409.81	−13.14%	−6.60%
Black Rail	1,601	51	3.21%	-88.86	-5.55%	-2.34%
Wood Stork	94,427	6,137	6.50%	-12409.76	-13.14%	-6.64%
Swallow-tailed Kite	189,642	1,783	0.94%	-34975.32	-18.44%	-17.50%
Bald Eagle	97,523	−1,703	-1.75%	-10490.24	-10.76%	-12.50%
Limpkin	6,002	-497	-8.28%	-100.64	-1.68%	-9.96%
Striped Newt	5,775	127	2.19%	-1516.65	-26.26%	-24.07%
American Oystercatcher	3444.4	6685.27	194.09%	-30.96	-0.90%	193.19%
Florida Scrub-Jay	32,269	-5,135	-15.91%	-5813.97	-18.02%	-33.93%
Bachman's Sparrow	51,105	-2,451	-4.80%	-9189.97	-17.98%	-22.78%
Round-tailed Muskrat	18,975	-170	-0.90%	-3081.04	-16.24%	-17.13%
Florida Mink	10,905	-2,686	-24.63%	-14.92	-0.14%	-24.77%
River Otter	184,616	-6,881	-3.73%	-32985.43	-17.87%	-21.59%
Florida Mouse	6,830	-380	-5.57%	-1342.05	-19.65%	-25.22%
Sherman's Fox Squirrel	51,770	-6,778	-13.09%	-7426.07	-14.34%	-27.44%
Florida Black Bear	169,642	871	0.51%	-34407.20	-20.28%	-19.77%
Sea Turtles	731.07	-495.81	-67.820%	-9.71	-1.328%	-69.148%
Shorebird Guild-Open water foraging	22238.7	33233.6	14.54%	-51.05	-0.23%	14.31%

Continued

Common Name	Current Habitat(Acres)	Net Gain/Loss to 0.5 m SLR	Percent Gain/ Loss to 0.5 m SLR	Loss to Land Use Change (Acres)	Percent Loss to Land Use Change	Total Percent Gain/Loss to 0.5 m SLR and Land Use Change
Shorebird Guild-Sand foraging	13205.69	5518.954	1.79%	-51.05	-0.39%	41.41%
Painted Bunting	9,463	-558	-5.90%	-2803.28	-29.62%	-35.52%
Anastasia Beach Mouse	1579.16	-23.87	-1.51%	-30.81	-1.95%	-3.46%
Merlin	11,846	2,209	18.65%	-15.86	-0.13%	18.51%
Atlantic Saltmarsh Snake	15,377	5,245	34.11%	-48.48	-0.32%	33.79%
Mangrove Forest Bird Guild	375.59	-4.25	-1.13%	-3.76	-1.00%	-2.13%
Worthington's Marsh Wren	6792.4084	1065.82	15.69%	-14.06	-0.21%	15.48%
Diamondback Terrapin	17,729	7,356	41.49%	-121.40	-0.68%	40.81%
MacGillivray's seaside Sparrow	10,905	-2,686	-24.63%	-14.92	0.14%	-24.77%

Continued

Common Name	Current Habitat(Acres)	Net Gain/Loss to 1.0 m SLR	Percent Gain/ Loss to 1.0 m SLR	Loss to Land Use Change (Acres)	Percent Loss to Land Use Change	Total Percent Gain/Loss to 1.0 m SLR and Land Use Change
Gopher Tortoise	13,346	−352	-2.60%	-2770.93	-20.76%	−23.40%
Spotted Turtle	104,397	−2,125	-2.00%	−18562.03	-17.78%	−19.82%
Eastern Indigo Snake	180,685	348	0.10%	-34119.77	-18.88%	-18.69%
Diamondback Rattle-snake	139,379	-316	-0.20%	-28081.98	-20.15%	-20.37%
Florida Kingsnake	42,407	347	0.80%	-7548.56	-17.80%	-16.98%
Florida Pine Snake	12,587	-446	-3.50%	-2653.76	-21.08%	-24.63%
Sandhill Crane	41,369	−1,129	-2.70%	-8279.00	-20.04%	-22.77%
Gopher Frog	68,501	-2,269	-3.30%	−10961.28	-20.44%	-23.76%
Southeastern American Kestrel	15,118	−1,800	−11.90%	−14003.26	-24.79%	-36.69%
Neotropical Migrant Forest Bird Guild	36,180	−1,559	-4.30%	-5207.14	−14.39%	−18.70%
Migratory (Wintering) Waterfowl	11,688	5,811	49.70%	−166.74	-1.43%	48.29%
Wading bird Guild	94,427	678	0.70%	−12295.28	−13.02%	−12.30%
Black Rail	1,601	−921	-58%	-5.49	-0.34%	-57.90%
Wood Stork	94,427	678	0.70%	−12295.35	−13.02%	−12.30%
Swallow-tailed Kite	189,642	-3,509	−1.80%	-34839.57	-18.37%	-20.22%
Bald Eagle	97,523	1,226	1.20%	−10389.07	−10.76%	-9.40%
Limpkin	6,002	-739	−12.30%	-468.18	−10.65%	-20.12%

Continued

Common Name	Current Habitat(Acres)	Net Gain/Loss to 1.0 m SLR	Percent Gain/ Loss to 1.0 m SLR	Loss to Land Use Change (Acres)	Percent Loss to Land Use Change	Total Percent Gain/Loss to 1.0 m SLR and Land Use Change
Striped Newt	5,775	222	3.80%	-1525.27	-26.41%	-22.58%
American Oystercatcher	3444.4	5777.22	167.73%	-40.70	-1.18%	166.55%
Florida Scrub-Jay	32,269	-2,180	-6.70%	-7009.73	-21.72%	-28.48%
Bachman's Sparrow	51,105	-2,101	-4.10%	-9223.92	-18.05%	-22.16%
Round-tailed Muskrat	18,975	-359	-1.80%	-3075.48	-16.21	-18.10%
Florida Mink	10,905	-4,948	-45.30%	-26.91	-0.25%	-45.62%
River Otter	184,616	-8,317	-4.50%	-32928.77	-17.84%	-22.34%
Florida Mouse	6,830	-218	-3.10%	-1363.03	-19.96%	-23.15%
Sherman's Fox Squirrel	51,770	-4,404	-8.50%	-7788.91	-15.05%	-23.55%
Florida Black Bear	169,642	-1,459	-0.86%	-31320.35	-18.46%	-19.32%
Sea Turtles	731.07	-469.47	-64.22	-9.59	-1.31%	-65.53%
Shorebird Guild- Open water foraging	22238.728	5678.36	25.53%	-77.00	-0.35%	25.19%
Shorebird Guild- Sand foraging	13205.691	-1406.77	-10.65%	-66.00	-0.50%	-11.15%
Painted Bunting	9,463	-501	-5.29%	-2817.95	-29.78%	-35.07%
Anastasia Beach Mouse	1579.167	-683.4	-43.28%	-35.83	-2.27%	-45.54%
Merlin	11,846	-1,396	-11.70%	-40.47	-0.34%	-12.12%
Atlantic Saltmarsh Snake	15,377	-46	-0.20%	-42.60	-0.28%	-0.58%

Continued

Common Name	Current Habitat(Acres)	Net Gain/Loss to 1.0 m SLR	Percent Gain/ Loss to 1.0 m SLR	Loss to Land Use Change (Acres)	Percent Loss to Land Use Change	Total Percent Gain/Loss to 1.0 m SLR and Land Use Change
Mangrove Forest Bird Guild	375.592	−8.75	−2.30%	−5.83	−1.55%	-3.88%
Worthington's Marsh Wren	6792.4084	−1241.01	−18.27%	−22.31	−0.33%	−18.60%
Diamondback Terrapin	17,729	6,337	35.70	−145.07	−0.82%	34.92%
MacGillivray's seaside Sparrow	10,905	−4,948	−45.30%	−26.91	−0.25%	−45.62%

Common Name	Current Habitat(Acres)	Net Gain/Loss to 2.5 m SLR	Percent Gain/ Loss to 2.5 m SLR	Loss to Land Use Change (Acres)	Percent Loss to Land Use Change	Total Percent Gain/Loss to 2.5 m SLR and Land Use Change
Gopher Tortoise	13,346	−3072.10	−23.02%	−2451.108	-18.37%	−41.39%
Spotted Turtle	104,397	−8424.38	−8.07%	−11063.69	−10.60%	−18.67%
Eastern Indigo Snake	180,685	−9730.5	−5.39%	−22119.107	−12.24	−17.63%
Diamondback Rattle-snake	139,379	−7570.25	−5.43%	−21544.09	−15.46%	−20.89%
Florida Kingsnake	42,407	−6780.15	−15.99	−4794.23	−11.31%	−27.29%
Florida Pine Snake	12,587	−3258.24	−25.89%	−2359.61	−18.75%	−44.63%
Sandhill Crane	41,369	3610.23	8.73%	−3860.81	−9.33%	−0.61%
Gopher Frog	68,501	−7008.72	−10.23%	−13004.4	−18.98%	−29.22%

Continued

Common Name	Current Habitat(Acres)	Net Gain/Loss to 2.5 m SLR	Percent Gain/ Loss to 2.5 m SLR	Loss to Land Use Change (Acres)	Percent Loss to Land Use Change	Total Percent Gain/Loss to 2.5 m SLR and Land Use Change
Southeastern American Kestrel	15,118	−3562.96	−23.57%	−3537.953	−23.40%	−46.97%
Neotropical Migrant Forest Bird Guild	36,180	−7731.17	−21.37%	−1039.45	−2.87%	−24.24%
Migratory (Wintering) Waterfowl	11,688	18620.86	159.32%	−4.72	−0.04%	159.28%
Wading bird Guild	94,427	−3340.17	−3.54%	−441.148	−0.47%	−4.00%
Black Rail	1,601	−994.65	−62.13%	0	0.00%	−62.13%
Wood Stork	94,427	−3339.83	−3.54%	−441.22	−0.47%	−4.00%
Swallow-tailed Kite	189,642	−14400.96	−7.59%	−22391.31	−11.81%	−19.40%
Bald Eagle	97,523	5868.77	6.02%	−3110.79	−3.19%	2.83%
Limpkin	6,002	−1746.03	−29.09%	0	0.00%	−29.09%
Striped Newt	5,775	535.44	9.27%	−1564.09	−27.09%	−17.81%
American Oystercatcher	3444.4	4226	122.69%	−47.1	−1.37%	121.32%
Florida Scrub-Jay	32,269	−9400.7	−29.13%	−6448.67	−19.98%	−49.12%
Bachman's Sparrow	51,105	−4130.42	−8.08%	−7942.66	−15.54%	−23.62%
Round−tailed Muskrat	18,975	−1636.3	−8.62%	−2146.93	−11.31%	−19.94%
Florida Mink	10,905	−5440.94	−49.90%	0	0.00%	−49.90%
River Otter	184,616	−16950.74	−9.18%	−21225.67	−11.50%	−20.68%

Continued

Common Name	Current Habitat(Acres)	Net Gain/Loss to 2.5 m SLR	Percent Gain/ Loss to 2.5 m SLR	Loss to Land Use Change (Acres)	Percent Loss to Land Use Change	Total Percent Gain/Loss to 2.5 m SLR and Land Use Change
Florida Mouse	6,830	−1308.49	−19.16	−1258.21	−18.42%	-37.58%
Sherman's Fox Squirrel	51,770	−8432.98	−16.29%	−5579.2	−10.78%	-27.07%
Florida Black Bear	169,642	−4526.16	−2.67%	19865	11.71%	-9.04%
Sea Turtles	731.07	−529.29	−72.40%	0	0	−72.40%
Shorebird Guild-Open water foraging	22238.728	17522.48	78.79%	−50.7	−0.23%	78.56%
Shorebird Guild- sand foraging	13205.691	−2344.5589	−17.75%	−11.64	−0.09%	−17.84%
Painted Bunting	9,463	−2151.97	−22.74%	−2648.22	−27.98%	−50.72%
Anastasia Beach Mouse	1579.167	−1387.17	−87.84%	0	0.00%	−87.84%
Merlin	11,846	−1862.59	−15.72%	0	0.00%	−15.72%
Atlantic Saltmarsh Snake	15,377	−2118.81	−13.78%0	0.00%	−13.78%	
Mangrove Forest Bird Guild	375.592	−374.7	−99.76%	0	0.00%	−99.76%
Worthington's Marsh Wren	6792.4084	−2268.11	−33.39%	−22.31	0.00%	−33.39%
Diamondback Terrapin	17,729	7159.18	40.38%	−119.67	−0.67%	−0.82%
MacGillivray's sea-sideSparrow	10,905	−5440.94	−49.90%	−26.91	0.00%	−49.90%

APPENDIX E
THE RUN SETTINGS FILE USED IN ZONATION

```
[Settings]
removal rule = 1
warp factor = 100
edge removal = 1
add edge points = 0
use SSI = 0
SSI file name = tutorial_input/SSI_list.txt
use planning unit layer = 0
planning unit layer file = tutorial_input/plu.asc

use cost = 1
cost file = tutorial_input/cost.tif
use mask = 1
mask file = tutorial_input/mask_towns.tif
use boundary quality penalty = 0
BQP profiles file = tutorial_input/BQPcurves.txt
BQP mode = 1
BLP = 0
use tree connectivity = 0
tree connectivity file = tutorial_input/tree.txt
use interactions = 0
interaction file = tutorial_input/interact.spp

annotate name = 0
logit space = 0
treat zero-areas as missing data = 0
z = 0.25
```

resample species = 0

[Info-gap settings]

Info-gap proportional = 0

use info-gap weights = 0

Info-gap weights file = tutorial_input/UCweights.spp

APPENDIX F
CROSSWALK BETWEEN SLAMM AND FLUCCS

SLAMM Code	SLAMM Description	FLUCCS Code
1	Developed Dry Land	Not used
2	Undeveloped Dry Land	Not used
3	Swamp	6300
4	Cypress Swamp	6210
5	Inland Fresh Marsh	6410
6	Tidal Fresh Marsh	6410
7	Scrub Shrub	6120
8	Regularly Flooded Marsh	6420
9	Mangrove	6120
10	Estuarine Beach	6510
11	Tidal Flat	6510
12	Ocean Beach	7100
13	Rocky Intertidal	6540
14	Inland Open Water	5000
15	Estuarine Water	5400
16	Tidal Creek	5400
17	Open Ocean	5700
18	Irregularly Flooded Marsh	6420
19	Vegetated Tidal Flat	6420

REFERENCES

Adams, V. M., Pressey, R. L., & Naidoo, R. (2010). Opportunity costs: Who really pays for conservation?. Biological Conservation, 143(2), 439-448.

Alvarez, R. (2012). The coastal build environment in Southeast Florida and sea level rise: Prognosis for potential damage and needed change. Panel Presentation: Sea Level Rise Summit 20-22 June 2012. Retrieved November 13, 2014 from http://mitigat.com/conferences/panel-presentation-sea-level-rise-summit-20-22-june-2012/.

Ando, A., Camm, J., Polasky, S., & Solow, A. (1998). Species distributions, land values, and efficient conservation. Science, 279(5359), 2126-2128.

Anon. (1992).National Forest Policy Statement: a New Focus for Australia's Forests. Australian Government Publishing Service, Canberra.

Araújo, M. B. (2009). Climate change and spatial conservation planning. Spatial conservation prioritization: quantitative methods and computational tools, 172-184.

Ball, I. R., Possingham, H. P., & Watts, M. (2009). Marxan and relatives: software for spatial conservation prioritisation. Spatial conservation prioritisation: quantitative methods and computational tools, 185-195.

Balmford, A., & Whitten, T. (2003). Who should pay for tropical conservation, and how could the costs be met? Oryx, 37(02), 238-250.

Bamber, J. L., Riva, R. E., Vermeersen, B. L., & LeBrocq, A. M. (2009). Reassessment of the potential sea-level rise from a collapse of the West Antarctic Ice Sheet.

Bekessy, S., & Gordon, A. (2007). Nurturing nature in the city. Steering sustainability in an urbanising world: Policy, practice and performance, 227-238.

Bengtsson, L., Hodges, K. I., & Keenlyside, N. (2009). Will extratropical storms intensify in a warmer climate?Journal of Climate, 22(9), 2276-2301Science, 324(5929), 901-903.

Box, G. E., & Draper, N. R. (1987). Empirical model-building and response surfaces. John Wiley & Sons.

Brinson, M. M., Christian, R. R., & Blum, L. K. (1995). Multiple states in the sea-level induced transition from terrestrial forest to estuary. Estuaries, 18(4), 648-659.

Burrough, P. A., McDonnell, R., Burrough, P. A., & McDonnell, R. (1998). Principles of geographical information systems (Vol. 333). Oxford: Oxford university press.

Callaway JC, DeLaune, RD, Patrick Jr WH (1997) Sediment accretion rates from four coastal wetlands along the Gulf of Mexico. J Coast Res 13(1):181-191.

Cameron D.S.E., J. R. Seavey, S. Claytor, T. Hoctor, M. Main, O. Mbuya, R. Noss, C. Rainyn. (2012). Florida Biodiversity Under a Changing Climate, Florida Climate Task Force. Retrieved June 10,2014 from http://floridaclimate.org/docs/biodiversity.pdf.

Church, J.A., P.U. Clark, A. Cazenave, J.M. Gregory, S. Jevrejeva, A. Levermann, M.A. Merrifield, G.A. Milne, R.S. Nerem, P.D. Nunn, A.J. Payne, W.T. Pfeffer, D. Stammer and A.S. Unnikrishnan(2013). Sea Level Change. In: Climate Change 2013: The Physical Science Basis. Contribution of Working Group I to the Fifth Assessment Report of the Intergovernmental Panel on Climate Change[Stocker, T.F., D. Qin, G.-K. Plattner, M. Tignor, S.K. Allen, J. Boschung, A. Nauels, Y. Xia, V. Bex and P.M. Midgley (eds.)]. Cambridge University Press, Cambridge, United Kingdom and New York, NY, USA.

Church, J. A., Clark, P. U., Cazenave, A., Gregory, J. M., Jevrejeva, S., Levermann, A., ... & Unnikrishnan, A. S. (2013). Sea-level rise by 2100. Science, 342(6165), 1445-1445.

Clough J S, Park RA, Fuller R (2010) SLAMM 6 beta technical documentation. Warren Pinnacle. Retrieved November 5, 2012 from http://warrenpinnacle.com/prof/SLAMM6/SLAMM6_Technical_Documentation.pdf.

Coastal States Organization (CSO) (2010). The faces of climate change adaptation: the need for proactive protection of the nation's coasts. Coastal State Organization, Climate Change Work Group, Washington.

Cox, J. (1994). Closing the gaps in Florida's wildlife habitat conservation system: recommendations to meet minimum conservation goals for declining wildlife species and rare plant and animal communities. Office of Environmental Services, FG & FWFC.

Cox, J. A., & Kautz, R. S. (2000). Habitat conservation needs of rare and imperiled wildlife in Florida. Office of Environmental Services, Florida Fish and Wildlife Conservation Commission.

Craft, C. (2007). Freshwater input structures soil properties, vertical accretion, and nutrient accumulation of Georgia and US tidal marshes. Limnology and Oceanography, 52(3), 1220-1230.

Delavenne, J., Metcalfe, K., Smith, R. J., Vaz, S., Martin, C. S., Dupuis, L., ... & Carpentier, A. (2012). Systematic conservation planning in the eastern English Channel:

comparing the Marxan and Zonation decision-support tools. ICES Journal of Marine Science: Journal du Conseil, 69(1), 75-83.

Desantis, L. R., Bhotika, S., Williams, K., & Putz, F. E. (2007). Sea-level rise and drought interactions accelerate forest decline on the Gulf Coast of Florida, USA. Global Change Biology, 13(11), 2349-2360.

Donnelly, J. P., & Bertness, M. D. (2001). Rapid shoreward encroachment of salt marsh cordgrass in response to accelerated sea-level rise. Proceedings of the National Academy of Sciences, 98(25), 14218-14223.

EPA (2014). Coastal Resiliency. Retrieved November 12, 2014 from http://water.epa.gov/infrastructure/greeninfrastructure/climate_res_coastal.cfm.

Ericson, J. P., Vörösmarty, C. J., Dingman, S. L., Ward, L. G., & Meybeck, M. (2006). Effective sea-level rise and deltas: causes of change and human dimension implications. Global and Planetary Change, 50(1), 63-82.

Feagin, R. A., Sherman, D. J., & Grant, W. E. (2005). Coastal erosion, global sea-level rise, and the loss of sand dune plant habitats. Frontiers in Ecology and the Environment, 3(7), 359-364.

Federal Geographic Data Committee. (1998). Geospatial Positioning Accuracy Standards, Part 3: National Standard for Spatial Data Accuracy. Subcommittee for Base Cartographic Data, 25p.

Ferrier, S., & Wintle, B. A. (2009). Quantitative approaches to spatial conservation prioritization: matching the solution to the need. Spatial conservation prioritization: Quantitative methods and computational tools, 1-15.

Florida Atlantic University (2008). Florida's Resilient Coasts: A State Policy Framework for Adaptation to Climate Change. Fort Lauderdale, Florida.

Florida Association of Counties (2013). Growth Management. Retrieved November 24, 2014 from http://www.fl-counties.com/advocacy/growth-management-agriculture-environment/growth-management.

Florida Department of Agriculture and Consumer Services (2013). Rural and Family Lands Protection Program. Retrieved November 24, 2014 from http://www.freshfromflorida.com/Divisions-Offices/Florida-Forest-Service/For-Landowners/Programs/Rural-and-Family-Lands-Protection-Program.

Florida Department of Environmental Protection (2011). Florida Lands and Waters– Brief Facts: Retrieved July 10, 2014 from http://www.dep.state.fl.us/lands/files/

FloridaNumbers_031011.pdf.

Florida Department of Environmental Protection (2012). Guana Tolomato Matanzas National Estuarine Research Reserve. Retrieved December 1, 2014 from http://www.dep.state.fl.us/gtm/.

Florida Fish and Wildlife Conservation Commission (2014). Florida's Cooperative Conservation Blueprint. Retrieved November 24, 2014 from http://myfwc.com/conservation/special-initiatives/blueprint/.

Galbraith, H., Jones, R., Park, R., Clough, J., Herrod-Julius, S., Harrington, B., & Page, G. (2002). Global climate change and sea level rise: potential losses of intertidal habitat for shorebirds. Waterbirds, 25(2), 173-183.

Geselbracht, L., Freeman, K., Kelly, E., Gordon, D. R., & Putz, F. E. (2011). Retrospective and prospective model simulations of sea level rise impacts on Gulf of Mexico coastal marshes and forests in Waccasassa Bay, Florida. Climatic Change, 107(1-2), 35-57.

Gesch, D., Oimoen, M., Greenlee, S., Nelson, C., Steuck, M., & Tyler, D. (2002). The national elevation dataset. Photogrammetric engineering and remote sensing, 68(1), 5-32.

Gesch, D. B. (2009). Analysis of lidar elevation data for improved identification and delineation of lands vulnerable to sea-level rise. Journal of Coastal Research, 49-58.

Gordon, A., Simondson, D., White, M., Moilanen, A., & Bekessy, S. A. (2009). Integrating conservation planning and landuse planning in urban landscapes. Landscape and urban planning, 91(4), 183-194.

GTM Research Reserve Management Plan (2009). Florida Department of Environmental Protection. Retrieved November 2, 2014 from http://www.dep.state.fl.us/gtm/.

Guisan, A., & Thuiller, W. (2005). Predicting species distribution: offering more than simple habitat models. Ecology letters, 8(9), 993-1009.

Hansen, J. E. (2007). Scientific reticence and sea level rise. Environmental research letters, 2(2), 024002.

Harris, L. D., & Cropper Jr, W. P. (1992). Between the devil and the deep blue sea: implications of climate change for Florida's fauna. Global Warming and Biological Diversity, 309-324.

Heller, N. E., & Zavaleta, E. S. (2009). Biodiversity management in the face of climate change: a review of 22 years of recommendations. Biological conservation, 142(1), 14-32.

Hijmans,R.J., and Elith, J. (2013). Species distribution modeling with R. Retrieved October 8, 2012 from http://cran.rproject.org/web/packages/dismo/vignettes/sdm.pdf.

Hoctor, T. S., Carr, M. H., & Zwick, P. D. (2000). Identifying a linked reserve system using a regional landscape approach: the Florida ecological network. Conservation Biology, 14(4), 984-1000.

Hoctor, T. S. (2011). Impacts Assessment Decision Support Model Project for the Southwest Florida Water Management District, Final Report. Southwest Florida Water Management District, Brooksville, FL.

Hoctor, T. S. (2003). Regional landscape analysis and reserve design to conserve Florida's biodiversity (Doctoral dissertation, University of Florida).

Hoctor, T.S., Volk, M., & Spontak, M. 2013. Updating the Florida Ecological Greenways Network. Final Report to the Florida Fish and Wildlife Conservation Commission.Retrieved August 10, 2014 from http://conservation.dcp.ufl.edu/Project-Downloads.html.

Hoctor, T., Oetting, J., Noss, R.F., Volk, M., Reece, J. (2014). Predicting and Mitigating the Impacts of Sea Level Rise and Land Use Change on Imperiled Species and Natural Communities. Final Report. Florida Fish and Wildlife Conservation Commission, Tallahassee, FL.

Hunter Jr, M. L. (1990). Coping with ignorance: the coarse-filter strategy for maintaining biodiversity. Balancing on the brink of extinction. Washington (DC): Island Press. p, 266-81.

Jantke K.(2011). Systematic Conservation Planning in Europe - the Case of Wetland Biodiversity. (Doctoral Dissertation, International Max Planck Research School on Earth System Modeling).

Kinser P, Curtis D, Beck B, Yates C (2007) Coastal wetlands of the Guana-Tolomato-Matanzas National Estuarine Research Reserve in Northeast Florida. St. Johns Water Management District, Palatka, FL.

Kukkala, A. S., & Moilanen, A. (2013). Core concepts of spatial prioritisation in systematic conservation planning. Biological Reviews, 88(2), 443-464.

Lauro, B., & Burger, J. (1989). Nest-site selection of American Oystercatchers (Haematopus palliatus) in salt marshes. The Auk, 185-192.

Lawler, J. J. (2009). Climate change adaptation strategies for resource management and conservation planning. Annals of the New York Academy of Sciences, 1162(1), 79-98.

Lehtomäki, J., Tomppo, E., Kuokkanen, P., Hanski, I., & Moilanen, A. (2009). Applying spatial conservation prioritization software and high-resolution GIS data to a national-scale study in forest conservation. Forest Ecology and Management, 258(11), 2439-2449.

Lehtomäki, J., & Moilanen, A. (2013). Methods and workflow for spatial conservation prioritization using Zonation. Environmental Modelling & Software, 47, 128-137.

Lindenmayer, D. B., Manning, A. D., Smith, P. L., Possingham, H. P., Fischer, J., Oliver, I., & McCarthy, M. A. (2002). The focal-species approach and landscape restoration: a critique. Conservation biology, 16(2), 338-345.

Linhoss, A. C., Kiker, G., Shirley, M., & Frank, K. (2014). Sea-Level Rise, Inundation, and Marsh Migration: Simulating Impacts on Developed Lands and Environmental Systems. Journal of Coastal Research.

Lowther, P. E., Lanyon, S. M., & Thompson, C. W. (1999). Painted Bunting(Passerina ciris). The Birds of North America, (398), 24.

Margules, C. R., & Pressey, R. L. (2000). Systematic conservation planning. Nature, 405(6783), 243-253.

Margules, C., Sarkar, S., & Margules, C. R. (2007). Systematic conservation planning. Cambridge University Press.

Maune, D. F., Maitra, J. B., & McKay, E. J. (2007). Accuracy standards and guidelines. Digital elevation model technologies and applications: The DEM users manual, 2, 65.

Mcleod, E., Poulter, B., Hinkel, J., Reyes, E., & Salm, R. (2010). Sea-level rise impact models and environmental conservation: A review of models and their applications. Ocean & Coastal Management, 53(9), 507-517.

Mitrovica, J. X., Gomez, N., & Clark, P. U. (2009). The sea-level fingerprint of West Antarctic collapse. Science, 323(5915), 753-753.

Moilanen, A., Franco, A. M., Early, R. I., Fox, R., Wintle, B., & Thomas, C. D. (2005). Prioritizing multiple-use landscapes for conservation: methods for large multi-species planning problems. Proceedings of the Royal Society B: Biological Sciences, 272(1575), 1885-1891.

Moilanen, A., & Wintle, B. A. (2006). Uncertainty analysis favours selection of spatially aggregated reserve networks. Biological Conservation, 129(3), 427-434.

Moilanen, A. (2007). Landscape zonation, benefit functions and target-based planning: unifying reserve selection strategies. Biological Conservation, 134(4), 571-579.

Moilanen, A., & Wintle, B. A. (2007). The Boundary-Quality Penalty: a Quantitative Method for Approximating Species Responses to Fragmentation in Reserve Selection. Conservation Biology, 21(2), 355-364.

Moilanen, A., Leathwick, J., & Elith, J. (2008). A method for spatial freshwater conservation prioritization. Freshwater Biology, 53(3), 577-592.

Moilanen, A., Kujala, H., & Leathwick, J. R. (2009). The Zonation framework and software for conservation prioritization. Spatial Conservation Prioritization, 196-210.

Moilanen, A., Wilson, K. A., & Possingham, H. P. (Eds.). (2009). Spatial conservation prioritization: quantitative methods and computational tools. Oxford, UK: Oxford University Press.

Moilanen, A., Anderson, B. J., Eigenbrod, F., Heinemeyer, A., Roy, D. B., Gillings, S., ... & Thomas, C. D. (2011). Balancing alternative land uses in conservation prioritization. Ecological Applications, 21(5), 1419-1426.

Molianen, A., L. Meller., J. Leppanen., F.M. Pouzols., A. Arponen., H. Kujala. (2012). Spatial conservation planning framework and software Zonation v3.1 user manual. Helsinki, Finland. Retried June 25, 2013 from http://cbig.it.helsinki.fi/software/zonation/.

Naidoo, R., & Ricketts, T. H. (2006). Mapping the economic costs and benefits of conservation. PLoS biology, 4(11), e360.

Naidoo, R., Balmford, A., Ferraro, P. J., Polasky, S., Ricketts, T. H., & Rouget, M. (2006). Integrating economic costs into conservation planning. Trends in Ecology & Evolution, 21(12), 681-687.

National Research Council (NRC) (2010). Adapting to the impacts of climate change. National Academy Press, Washington.

Nicholls, R. J., & Cazenave, A. (2010). Sea-level rise and its impact on coastal zones. science, 328(5985), 1517-1520.

NOAA (2014). Sea level affecting marshes model. Retrieved November 18, 2014 from http://coast.noaa.gov/digitalcoast/tools/slamm.

Noss, R. F. (2011). Between the devil and the deep blue sea: Florida's unenviable position with respect to sea level rise. Climatic Change, 107(1-2), 1-16.

Noss, R. F., Reece, J.S., Hoctor, T., Volk, M., Oetting, J. (2014). Adaptation to Sea-level rise in Florida: Biological Conservation Priorities . Final Report. To the Kresge Foundation. Retrieved September 10, 2014 from http://conservation.dcp.ufl.edu/Project-Downloads.html.

NWF (2006) An Unfavorable tide - global warming, coastal habitats and sportfishing in Florida. National Wildlife Federation, Florida Wildlife Federation.

Nyman, J. A., DeLaune, R. D., Roberts, H. H., & Patrick Jr, W. H. (1993). Relationship between vegetation and soil formation in a rapidly submerging coastal marsh. Marine ecology progress series. Oldendorf, 96(3), 269-279.

Pagiola, S., Landell-Mills, N., & Bishop, J. (2002). Market-based mechanisms for forest conservation and development. Selling Forest Environmental Services. Market-based Mechanisms for Conservation and Development, 1-13.

Parmesan, C., & Yohe, G. (2003). A globally coherent fingerprint of climate change impacts across natural systems. Nature, 421(6918), 37-42.

Pearson, R.G. 2007. Species' Distribution Modeling for Conservation Educators and Practitioners.Synthesis. American Museum of Natural History. Retrieved November 5, 2013 from http://ncep.amnh.org.

Peters, R. L., & Darling, J. D. (1985). The greenhouse effect and nature reserves. Bioscience, 707-717.

Pilkey, O. H., & Young, R. (2008). The rising sea. Island Press.Washington.

Polasky, S., Camm, J. D., & Garber-Yonts, B. (2001). Selecting biological reserves cost-effectively: an application to terrestrial vertebrate conservation in Oregon. Land Economics, 77(1), 68-78.

Possingham, H., Ball, I., & Andelman, S. (2000). Mathematical methods for identifying representative reserve networks. In Quantitative methods for conservation biology (pp. 291-306). Springer New York.

Poulter, B., & Halpin, P. N. (2008). Raster modelling of coastal flooding from sea-level rise. International Journal of Geographical Information Science, 22(2), 167-182.

Pressey, R. L., Cabeza, M., Watts, M. E., Cowling, R. M., & Wilson, K. A. (2007). Conservation planning in a changing world. Trends in ecology & evolution, 22(11), 583-592.

Pressey, R. L., & Bottrill, M. C. (2008). Opportunism, threats, and the evolution of systematic conservation planning. Conservation Biology, 22(5), 1340-1345.

Pressey, R. L., Watts, M. E., Barrett, T. W., & Ridges, M. J. (2009). The C-Plan conservation planning system: origins, applications, and possible futures. Spatial conservation prioritization: quantitative methods and computational tools, 211-34.

Redford, K. H., Coppolillo, P., Sanderson, E. W., Da Fonseca, G. A., Dinerstein, E.,

Groves, C., ... & Wright, M. (2003). Mapping the conservation landscape. Conservation biology, 17(1), 116-131.

Regan, H. M., Ensbey, M. J., & Burgman, M. A. (2009). Conservation prioritization and uncertainty in planning inputs. Oxford University Press, 328pp.

Robert, S. (1994). Peninsular Malaysia's protected areas' coverage, 1903-92: creation, rescission, excision, and intrusion. Environmental Conservation, 21(1).

Root, T. L., Price, J. T., Hall, K. R., Schneider, S. H., Rosenzweig, C., & Pounds, J. A. (2003). Fingerprints of global warming on wild animals and plants. Nature, 421(6918), 57-60.

Runting, R. K., Wilson, K. A., & Rhodes, J. R. (2013). Does more mean less? The value of information for conservation planning under sea level rise. Global change biology, 19(2), 352-363.

Sala, O. E., Chapin, F. S., Armesto, J. J., Berlow, E., Bloomfield, J., Dirzo, R., ... & Wall, D. H. (2000). Global biodiversity scenarios for the year 2100. science, 287(5459), 1770-1774.

Simberloff, D. (1998). Flagships, umbrellas, and keystones: is single-species management passé in the landscape era?. Biological conservation, 83(3), 247-257.

Small, C., & Nicholls, R. J. (2003). A global analysis of human settlement in coastal zones. Journal of Coastal Research, 584-599.

Solomon, S. (Ed.). (2007). Climate change 2007-the physical science basis: Working group I contribution to the fourth assessment report of the IPCC (Vol. 4). Cambridge University Press.

Tallis, H., Kareiva, P., Marvier, M., & Chang, A. (2008). An ecosystem services framework to support both practical conservation and economic development. Proceedings of the National Academy of Sciences, 105(28), 9457-9464.

Tallis, H.T., Ricketts, T., Guerry, A.D., Wood, S.A., Sharp, R., Nelson, E., Ennaanay, D., Wolny, S., Olwero, N., Vigerstol, K., Pennington, D., Mendoza, G., Aukema, J., Foster, J., Forrest, J., Cameron, D., Arkema, K., Lonsdorf, E., Kennedy, C., Verutes, G., Kim, C.K., Guannel, G., Papenfus, M., Toft, J., Marsik, M., Bernhardt, J., and Griffin, R., Glowinski, K., Chaumont, N., Perelman, A., Lacayo, M. Mandle, L., Griffin, R., Hamel, P., Chaplin-Kramer, R. (2013). InVEST 2.6.0 User's Guide. The Natural Capital Project, Stanford.

Titus, J. G., & Cacela, D. (2008). Uncertainty ranges associated with EPA's estimates of the area of land close to sea level. Background documents supporting climate change science program synthesis and assessment product, 4.

Turner, W. R., Bradley, B. A., Estes, L. D., Hole, D. G., Oppenheimer, M., & Wilcove, D. S. (2010). Climate change: helping nature survive the human response. Conservation Letters, 3(5), 304-312.

UN Atlas of the Oceans (2010). Human settlements on the Coast. Retrieved December 1,2014 from http://www.oceansatlas.org/servlet/CDSServlet?status=ND0xODc3JjY9ZW4mMzM9KiYzNz1rb3M~.

United States Census Bureau(2010). U.S. 2010 Census Data. Retrieved July 10, 2014 from http://www.census.gov/2010census/data/.

United States Department of Agriculture. (2012). Wetlands Reserve Program. Retrieved November 24,2014 from http://www.nrcs.usda.gov/wps/portal/nrcs/main/national/programs/easements/wetlands/.

United States Fish & Wildlife Services (2010). The Peninsular Florida Landscape Cooperative. Retrieved November 24, 2014 from www.fws.gov/southeast/LCC/.../FlordiaLCCFactSheet.pdf.

United States Fish & Wildlife Services (2014). Landscape Conservation Cooperatives. Retrieved November 24, 2014 from http://www.fws.gov/landscapeconservation/lcc.html.

Waetzold, F., Drechsler, M., Armstrong, C. W., Baumgaertner, S., Grimm, V., Huth, A., ... & Wissel, C. (2006). Ecological-Economic Modeling for Biodiversity Management: Potential, Pitfalls, and Prospects. Conservation Biology, 20(4), 1034-1041.

Westerling, A. L., Hidalgo, H. G., Cayan, D. R., & Swetnam, T. W. (2006). Warming and earlier spring increase western US forest wildfire activity. science, 313(5789), 940-943.

Wetzel, F. T., Kissling, W. D., Beissmann, H., & Penn, D. J. (2012). Future climate change driven sea-level rise: secondary consequences from human displacement for island biodiversity. Global change biology, 18(9), 2707-2719.

Wilson SG, Fischetti TR (2010). Coastline population trends in the United States: 1960 to 2008. US Census Bureau, Washington. Retrieved August 10, 2014 from www.census.gov/prod/2010pubs/p25-1139.pdf.

Wise, S. M. (1998). The effect of GIS interpolation errors on the use of digital elevation models in geomorphology. Landform monitoring, modelling and analysis, 99, 139-164.

Zwick, D. P., & Carr. H. M. (2006). Florida 2060: A Population Distribution Scenario for the State of Florida. A research project prepared for 1000 Friends of Florida. Retrieved August 9, 2014 from http://www.1000friendsofflorida.org/connecting-people/florida2060/.

FIGURES

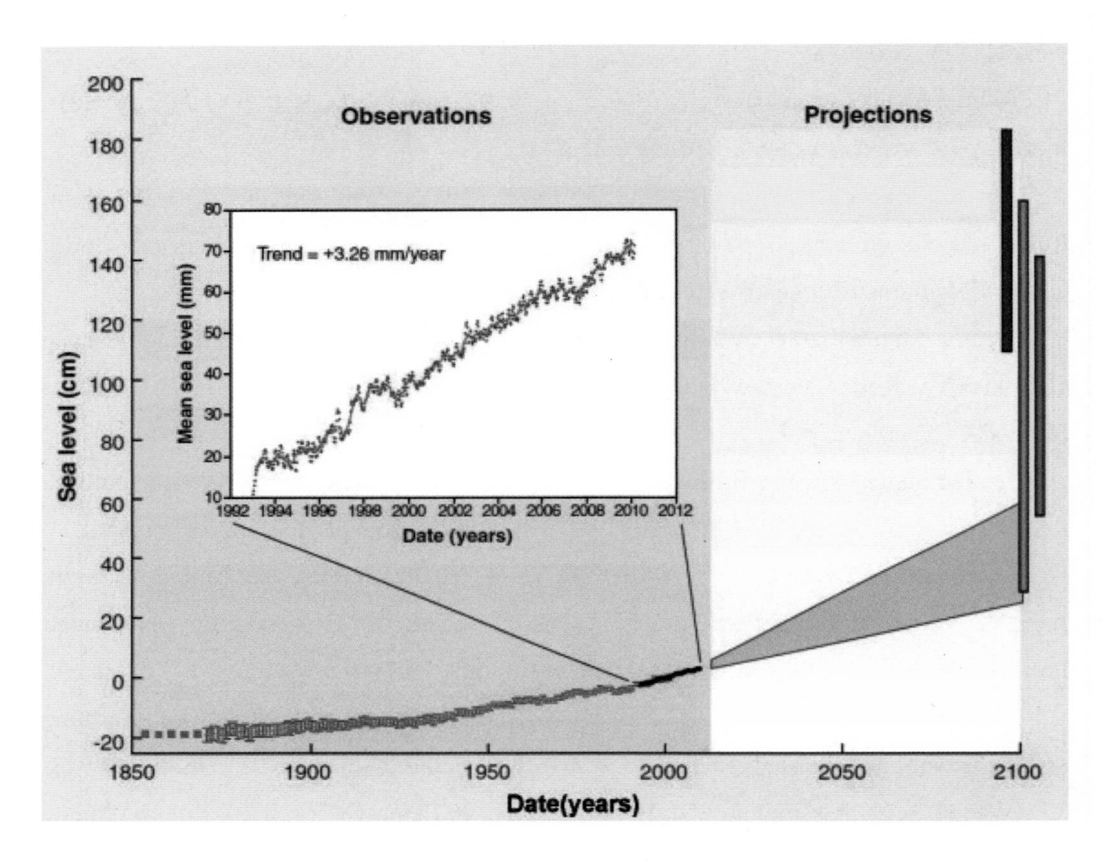

Figure 1-1 "Global mean sea level evolution over the 20th and 21st centuries

The red curve is based on tide gauge measurements. The black curve is the altimetry record (zoomed over the 1993–2009 time span). Projections for the 21st century are also shown. The shaded light blue zone represents IPCC 4th assessment reportprojections for the A1FI greenhouse gas emission scenario. (Nicholls & Cazenave, 2010)". (Source: Cameron et al., 2012)

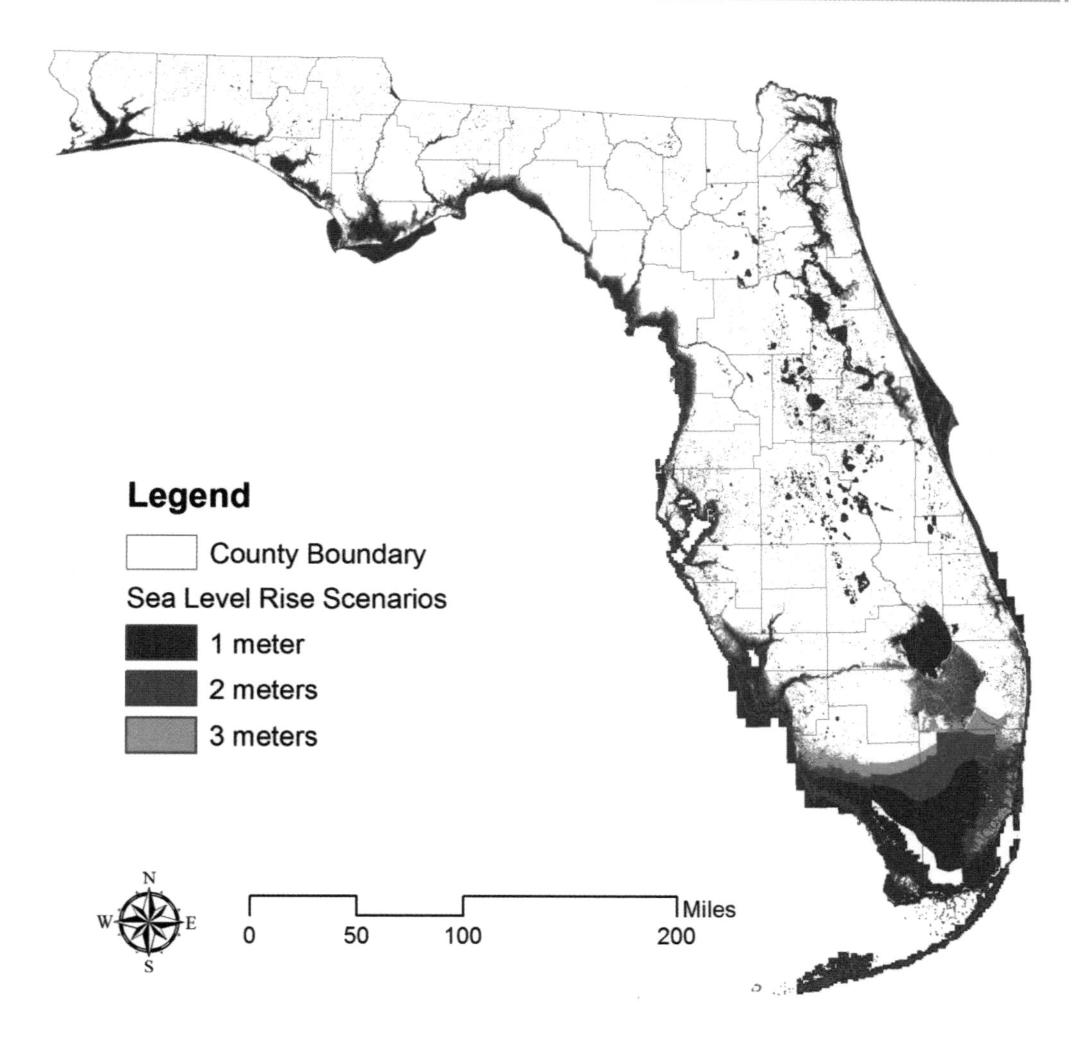

Figure 1-2 Sea level rise scenarios in Florida based on 5 meter cell size digital elevation model (DEM) from 4 different sources, with the following order of priority: 1) Northwest Florida Water Management District Digital Elevation Model (DEM), 2) FLIDAR Coastal DEM, 3) Statewide Fish & Wildlife Conservation Commission DEM, 4) Contour Derived DEM

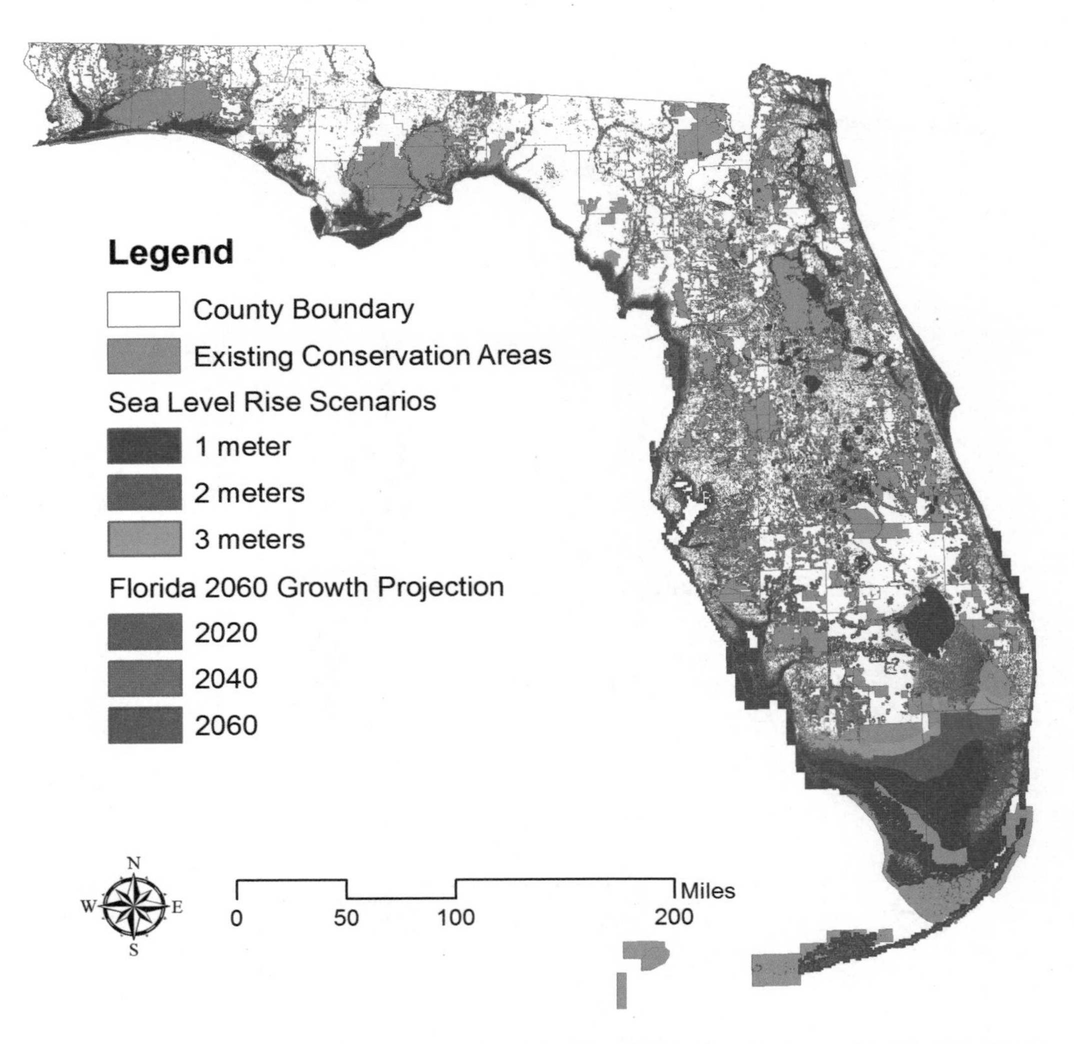

Figure 1-3 Existing conservation areas are under direct threat from SLR ,and future population growth, shift and increasing urbanization will replace potential areas for future conservation in Florida

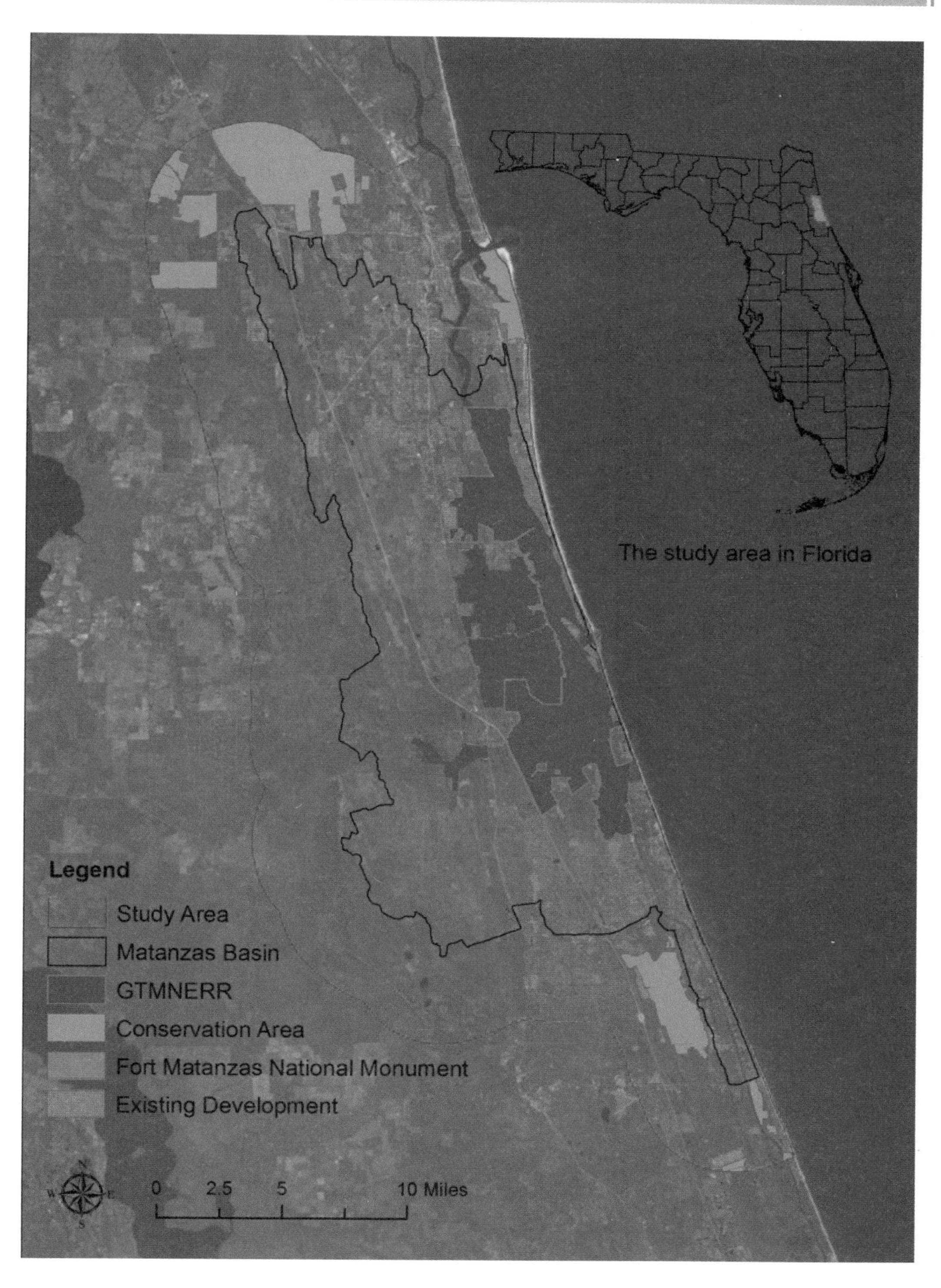

Figure 1-4 The Matanzas study area includes a 5 km buffer beyond the Matanzas River Basin in Northeast Florida to include regional ecological considerations

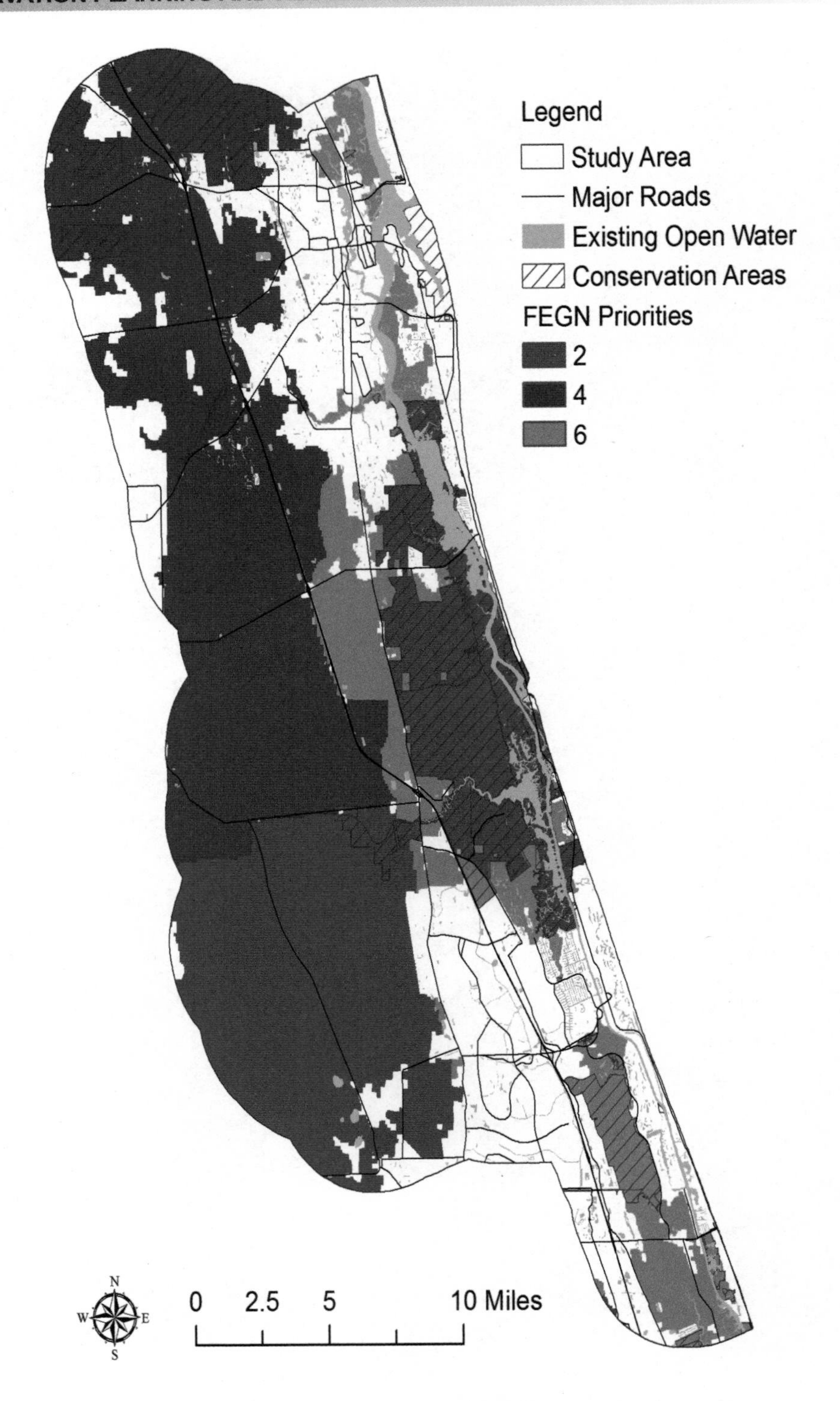

Figure 1-5 Florida Ecological Greenways Network identified in the Matanzas study area

Figure 1-6 Critical Lands and Waters Identification Project Dataset identified in the Matanzas study area

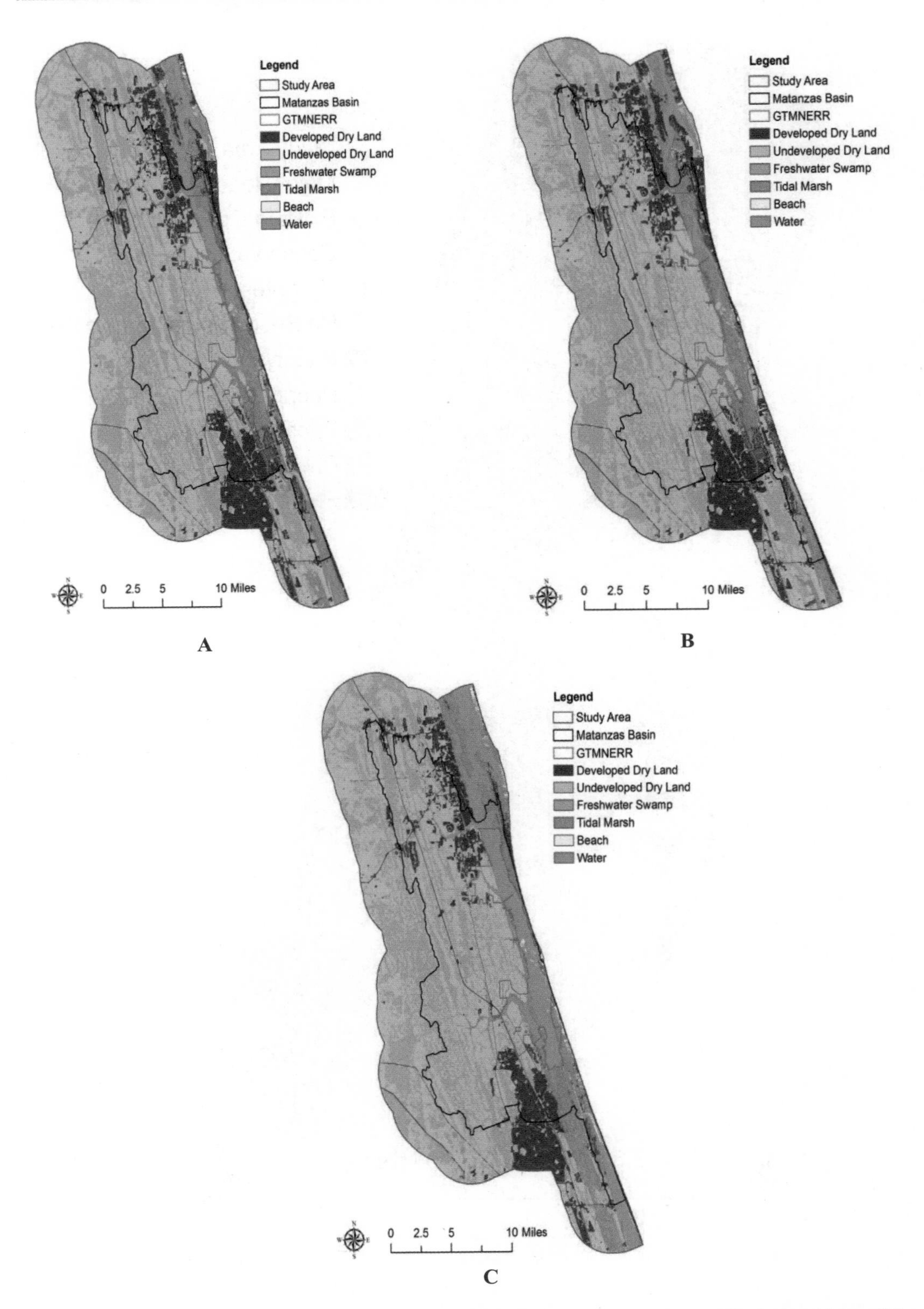

Figure 2-1 Maps of SLAMM results showing coastal habitat change by 2100, derived from (A) 0.5 m SLR, (B) 1.0 m SLR and (C) 2.5 m SLR

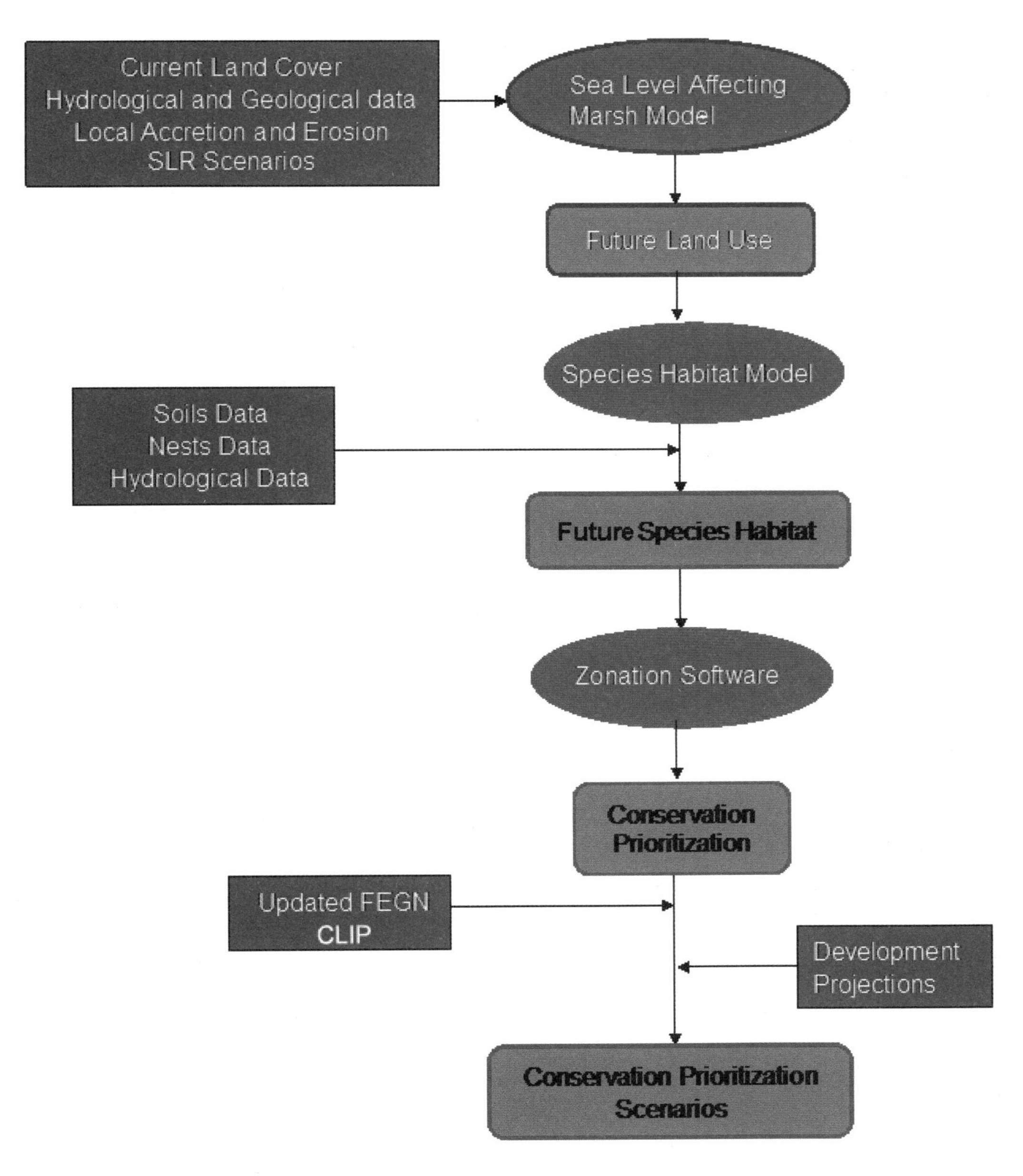

Figure 2-2 The integrated modeling framework used to identify SLR adaptive conservation prioritization

The blue oval components are the three models used in the planning process. The purple rectangle components are data inputs and the green components are primary outputs of each analysis

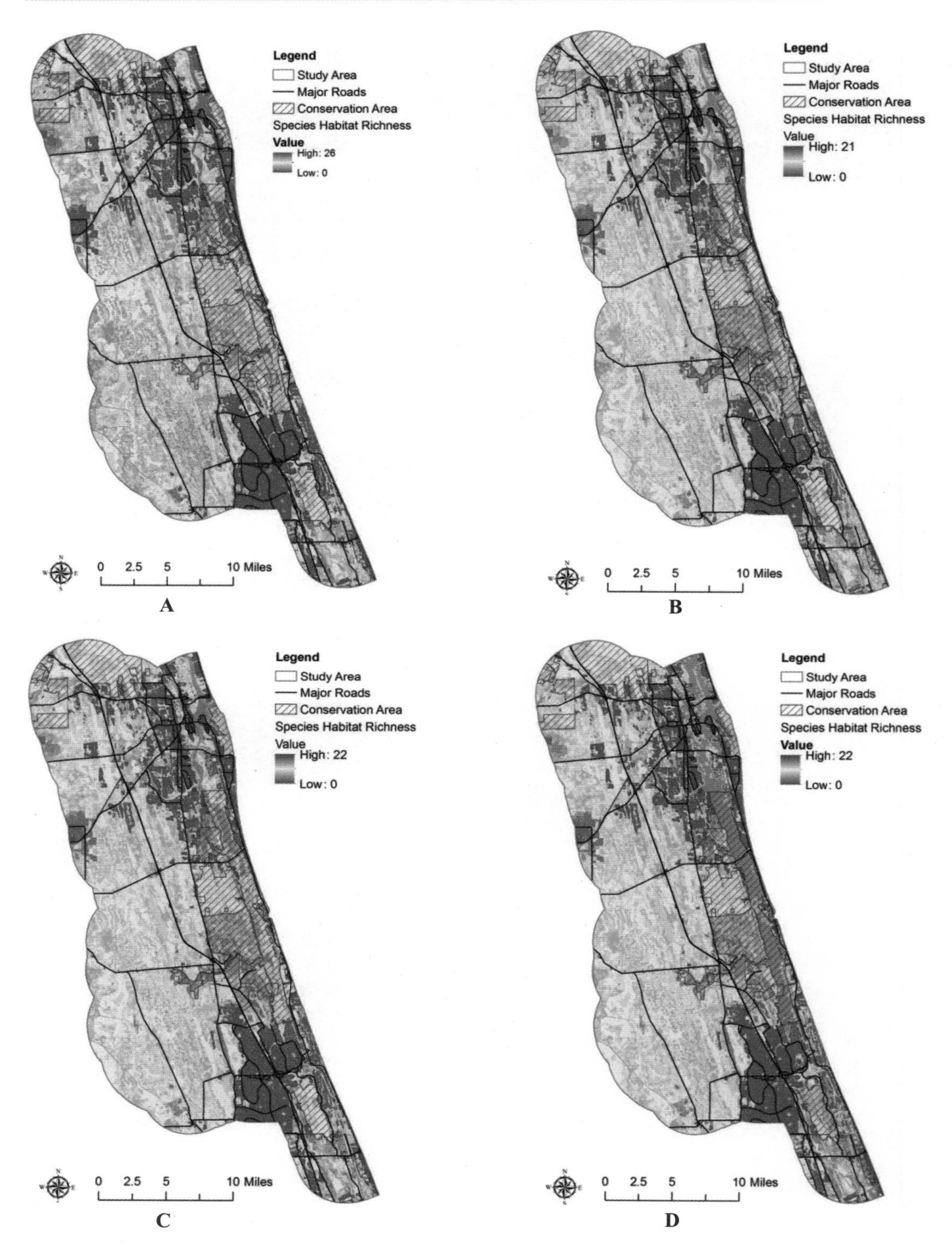

Figure 2-3 Species habitat richness under (A) current condition (B) 0.5 m SLR, (C) 1.0 m SLR and (D) 2.5 m SLR

The blue color represents high species habitat richness areas

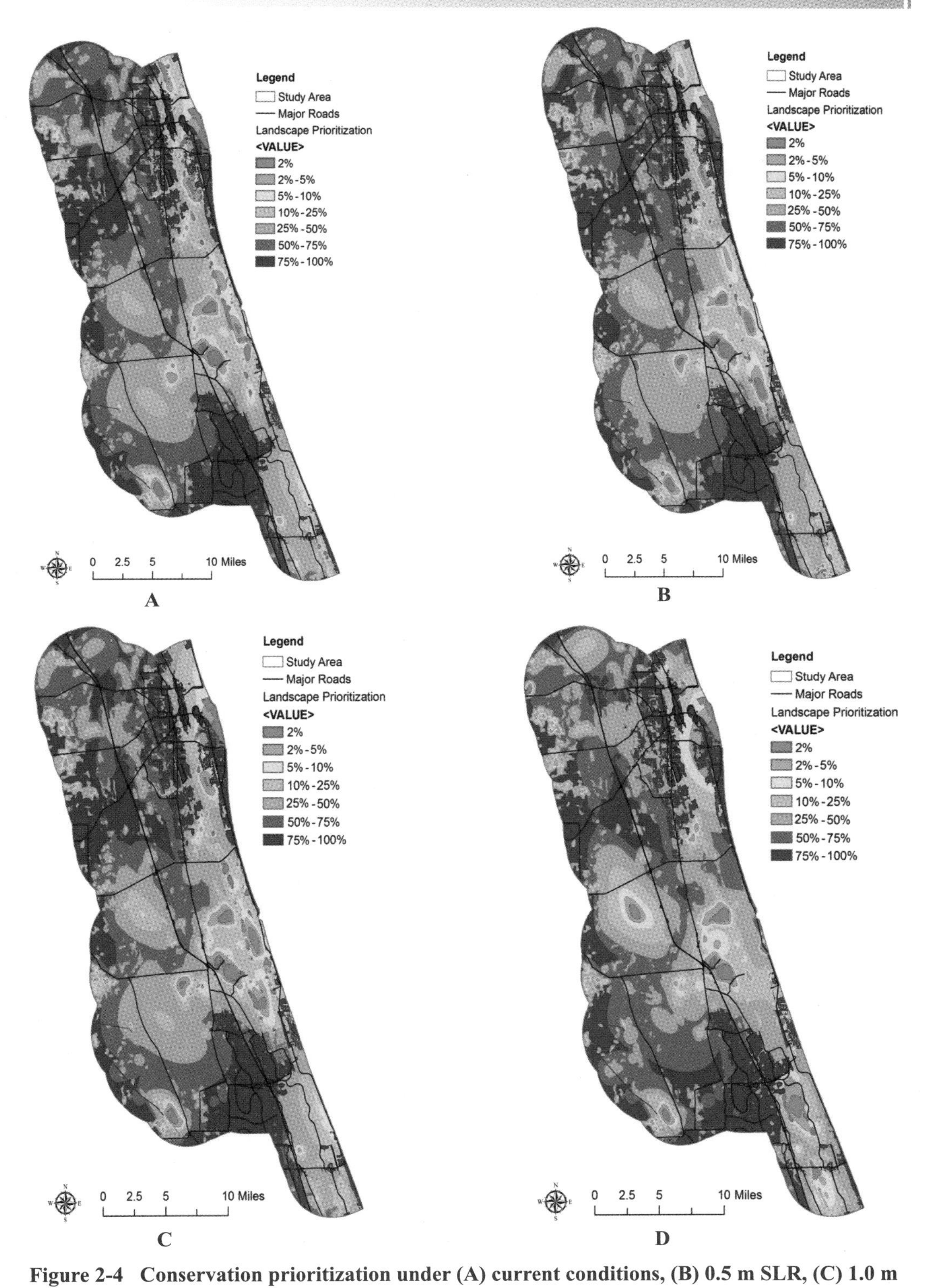

Figure 2-4 Conservation prioritization under (A) current conditions, (B) 0.5 m SLR, (C) 1.0 m SLR and (D) 2.5 m SLR

The red color represents high conservation priorities and the blue color represents low conservation priorities

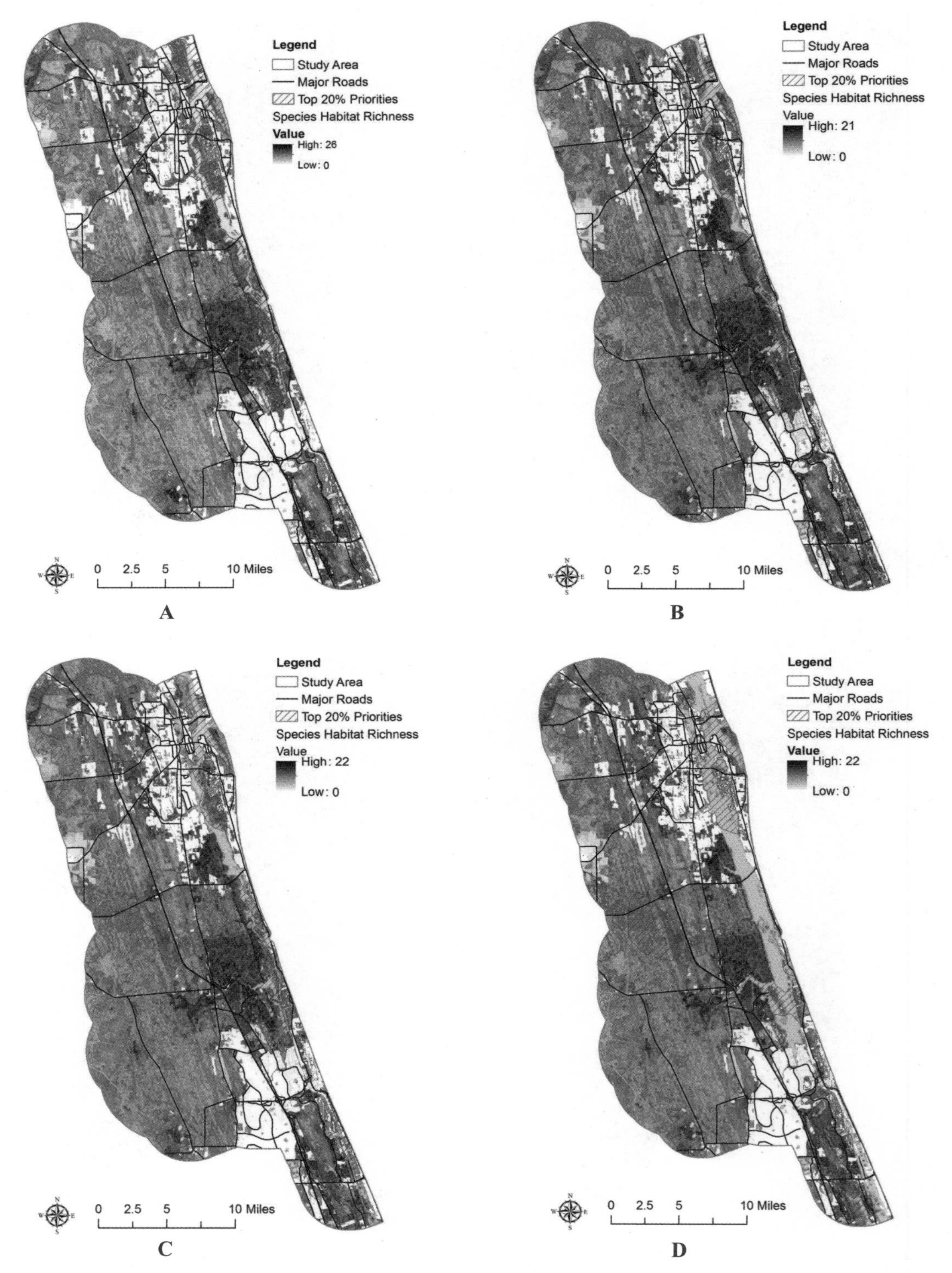

Figure 2-5 Top 20% landscape priorities with species habitat richness under (A) current condition, (B) 0.5 m SLR, (C) 1.0 m SLR and (D) 2.5 m SLR

The red hatching represents the top 20% conservation priority areas identified by Zonation

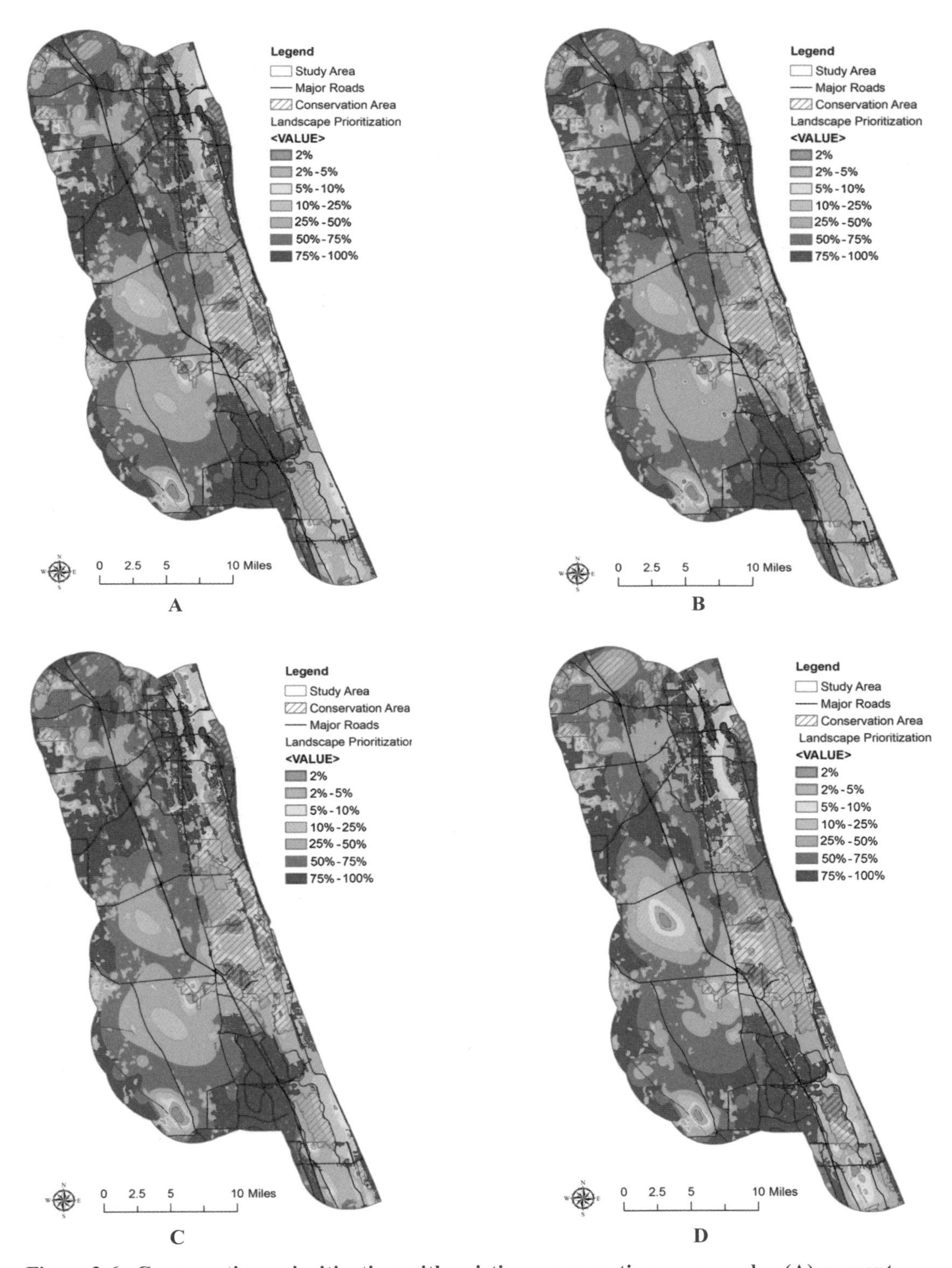

Figure 2-6 Conservation prioritization with existing conservation areas under (A) current condition, (B) 0.5 m SLR, (C) 1.0 m SLR and (D) 2.5 m SLR

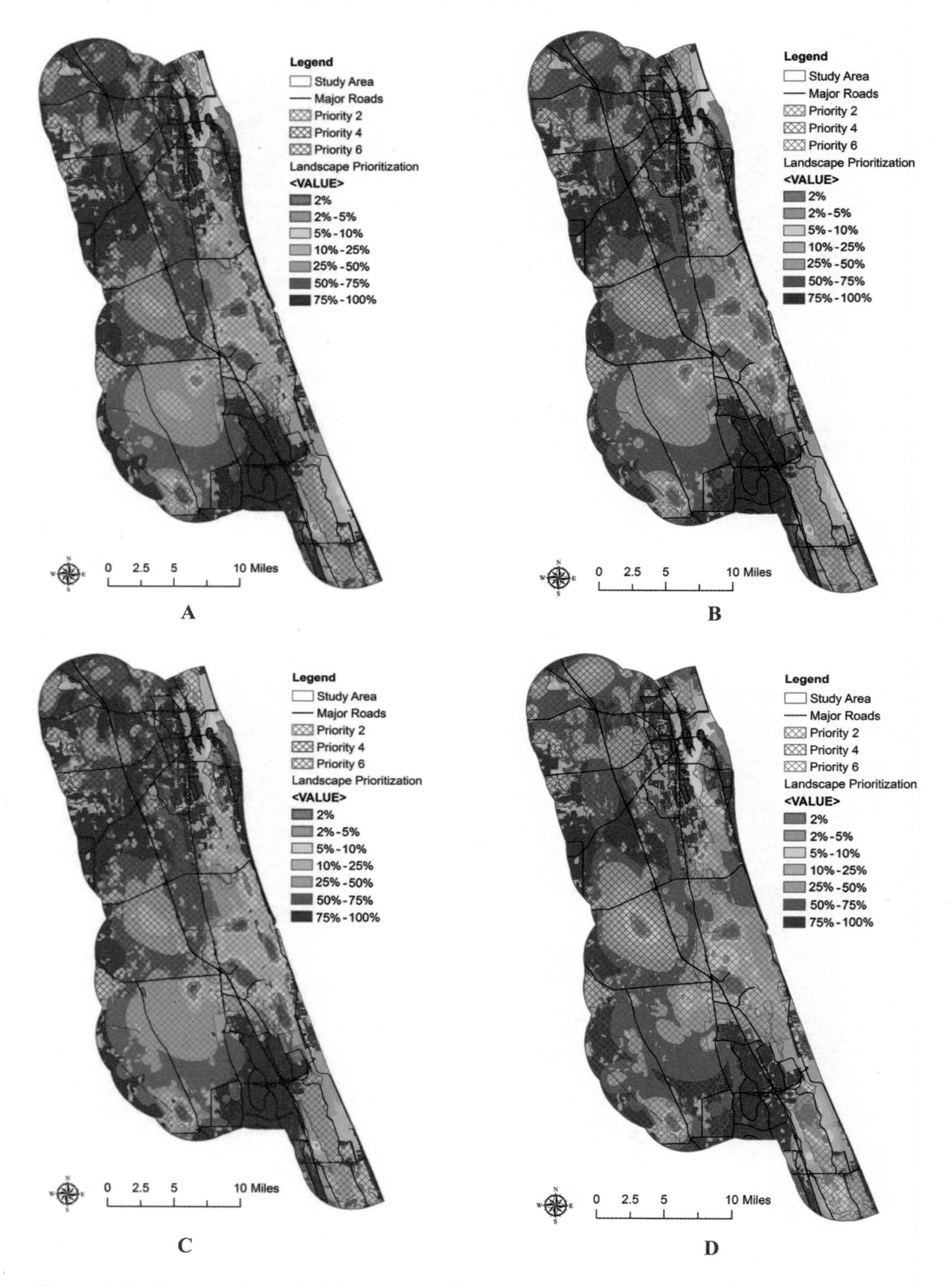

Figure 2-7 Conservation prioritization with FEGN under (A) current condition, (B) 0.5 m SLR, (C) 1.0 m SLR and (D) 2.5 m SLR
Priority 2 is the highest FEGN priority in the study area

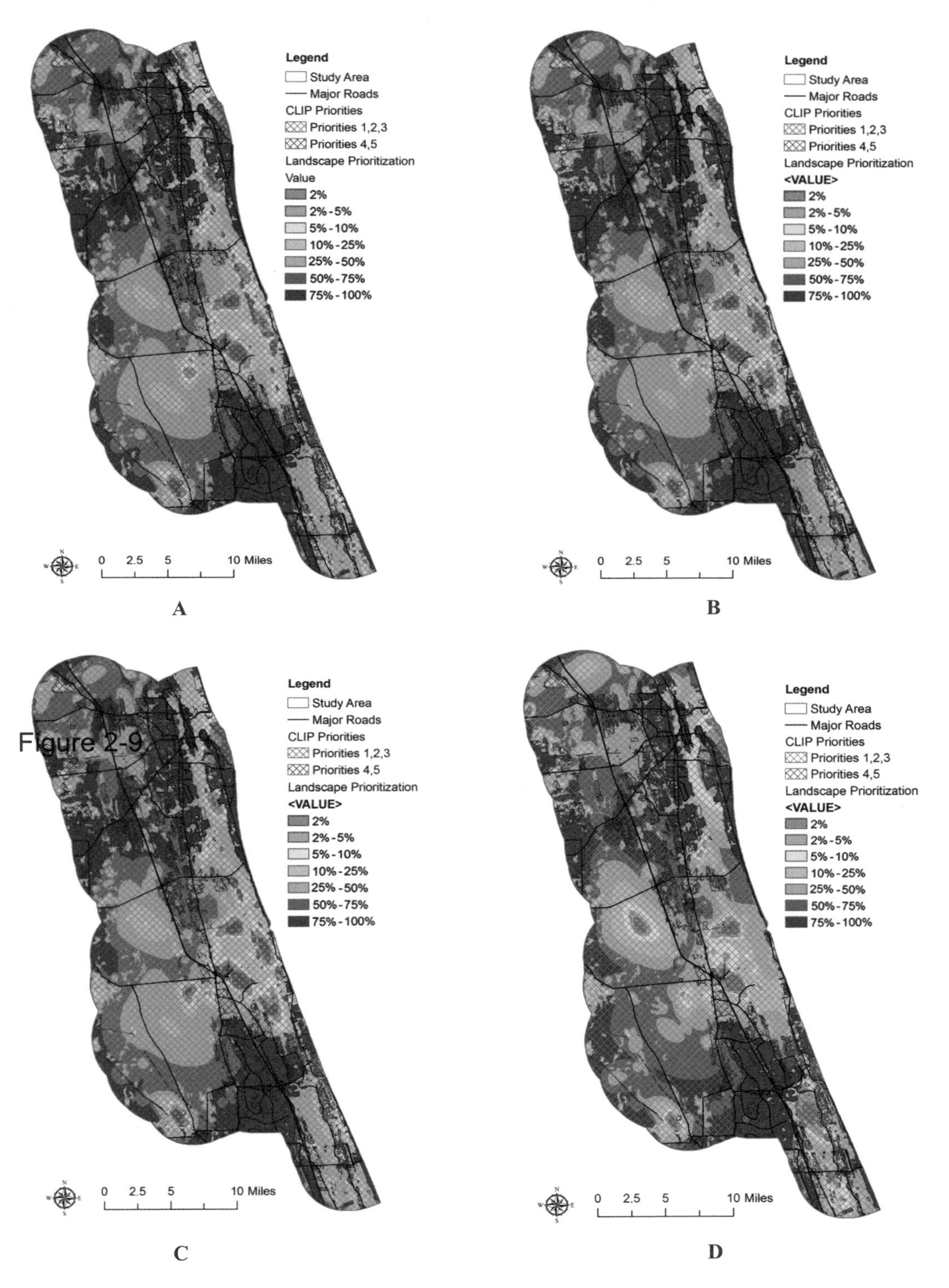

Figure 2-8 Conservation prioritization with the CLIP 3.0 as cross hatching under (A) current conditions, (B) 0.5 m SLR, (C) 1.0 m SLR and (D) 2.5 m SLR

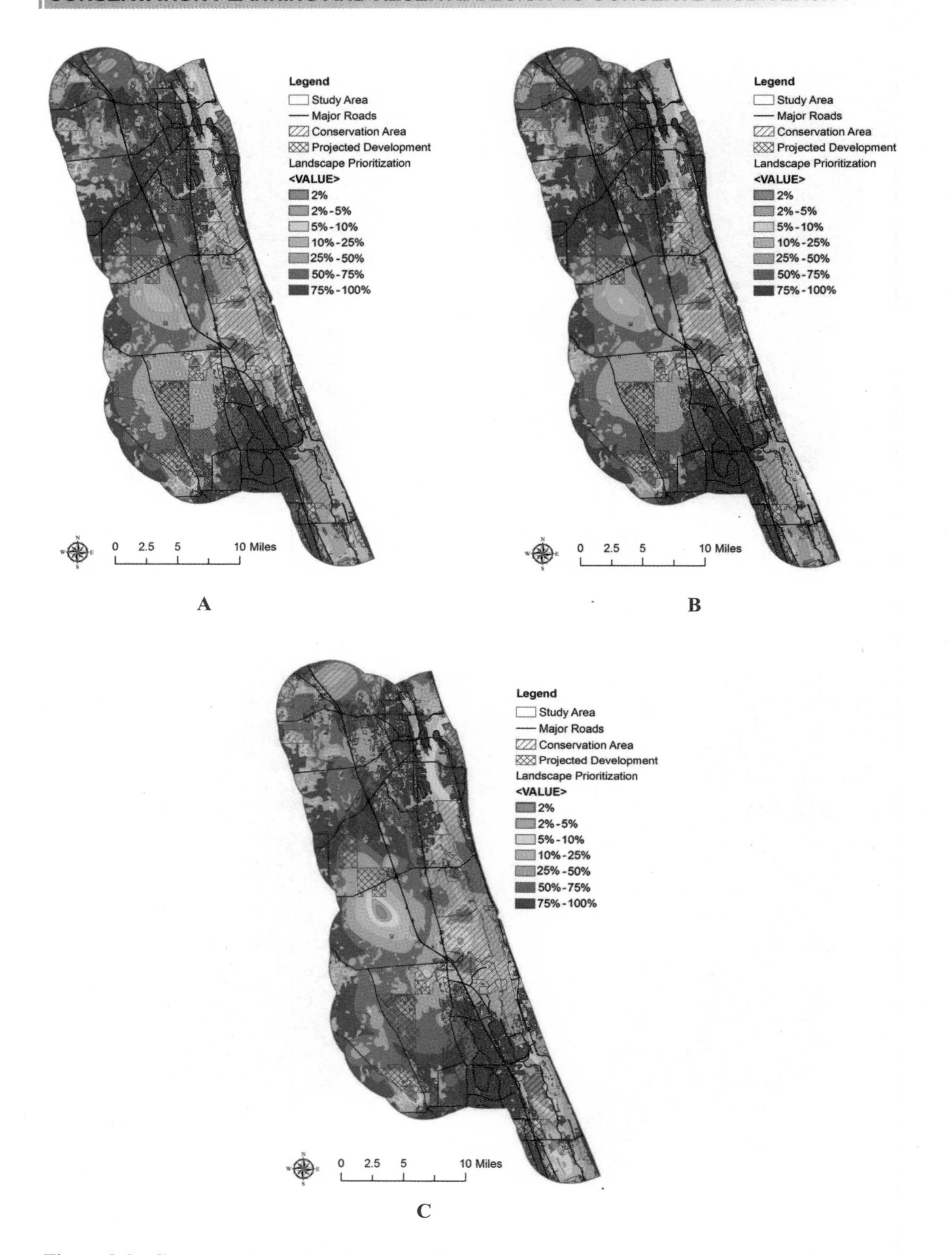

Figure 2-9 Conservation prioritization with the projected development as cross hatching under (A) 0.5 m SLR, (B) 1.0 m SLR and (C) 2.5 m SLR
The red color represents top conservation prioritiesin the study area

Figure 2-10 The change of top 10% conservation priorities for the current conditions and the 0.5 m, 1.0 m and 2.5 m SLR scenarios

The dark blue color represents top 10% conservation priorities under all SLR scenarios

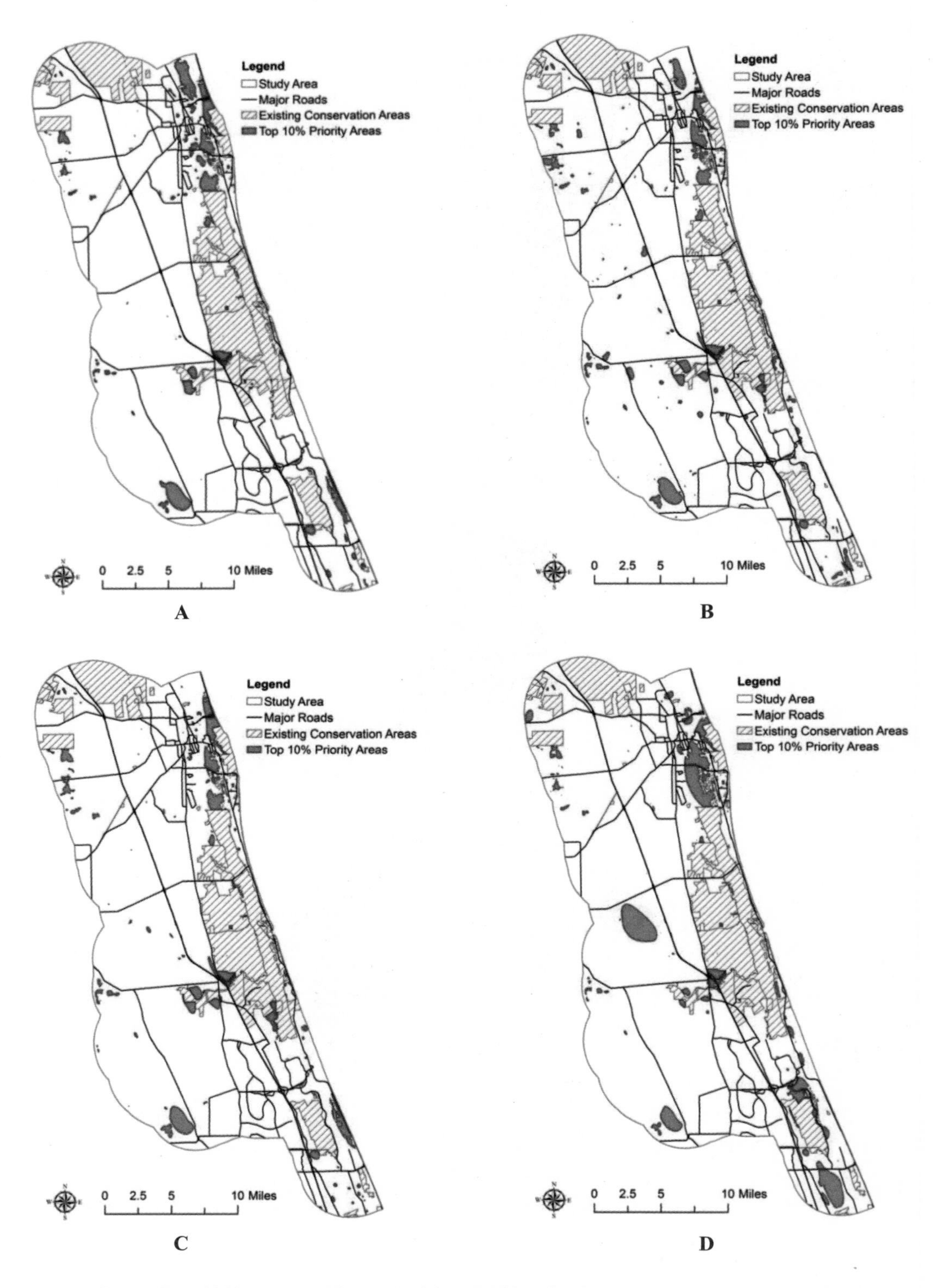

Figure 2-11 Top 10% conservation priorities outside of existing conservation areas under (A) current condition, (B) 0.5 m SLR, (C) 1.0 m SLR and (D) 2.5 m SLR

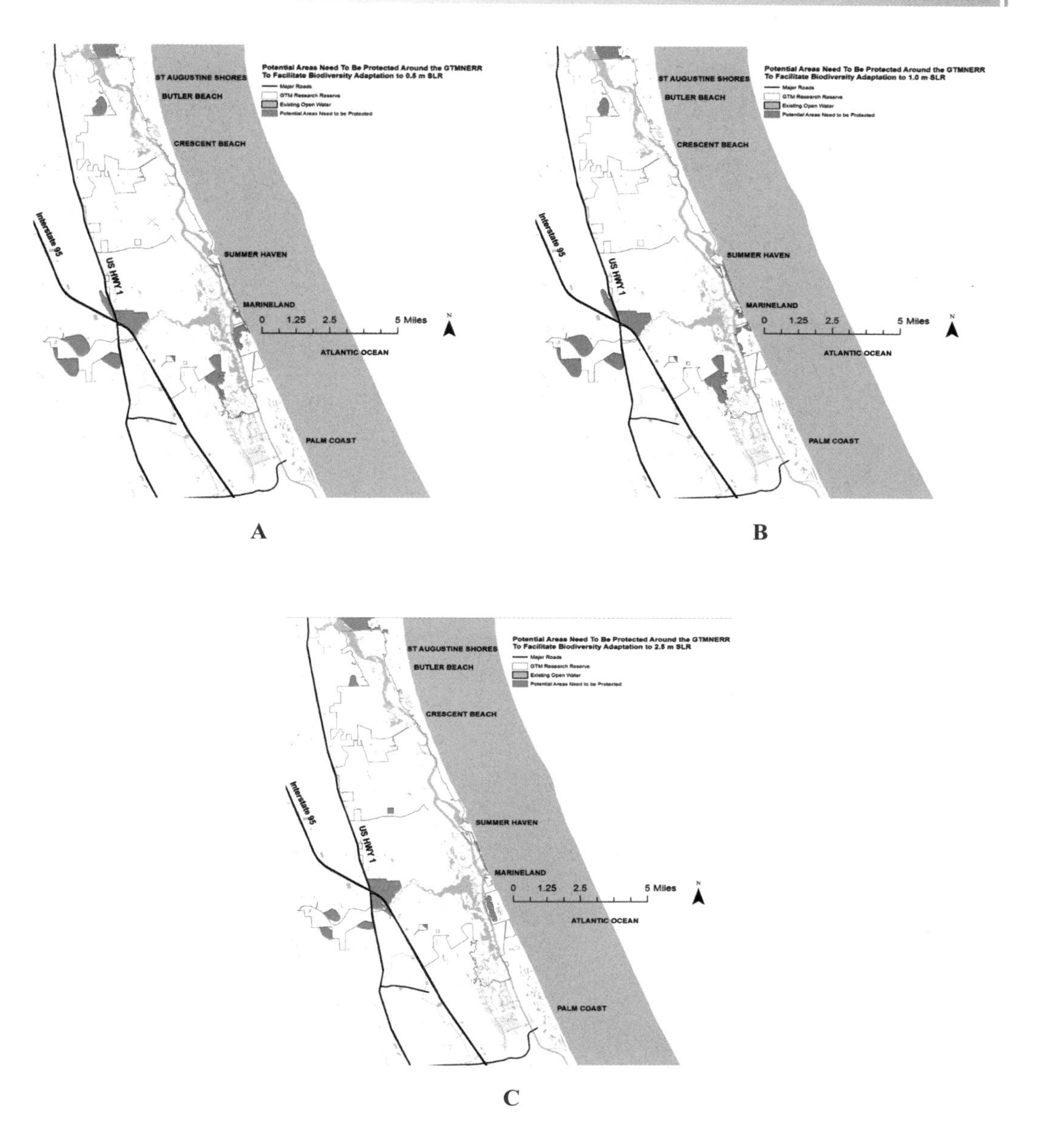

Figure 2-12 Potential areas need to be protected to facilitate biodiversity adaptation to SLR around GTMNERR under (A) 0.5 m SLR, (B) 1.0 m SLR and (C) 2.5 m SLR

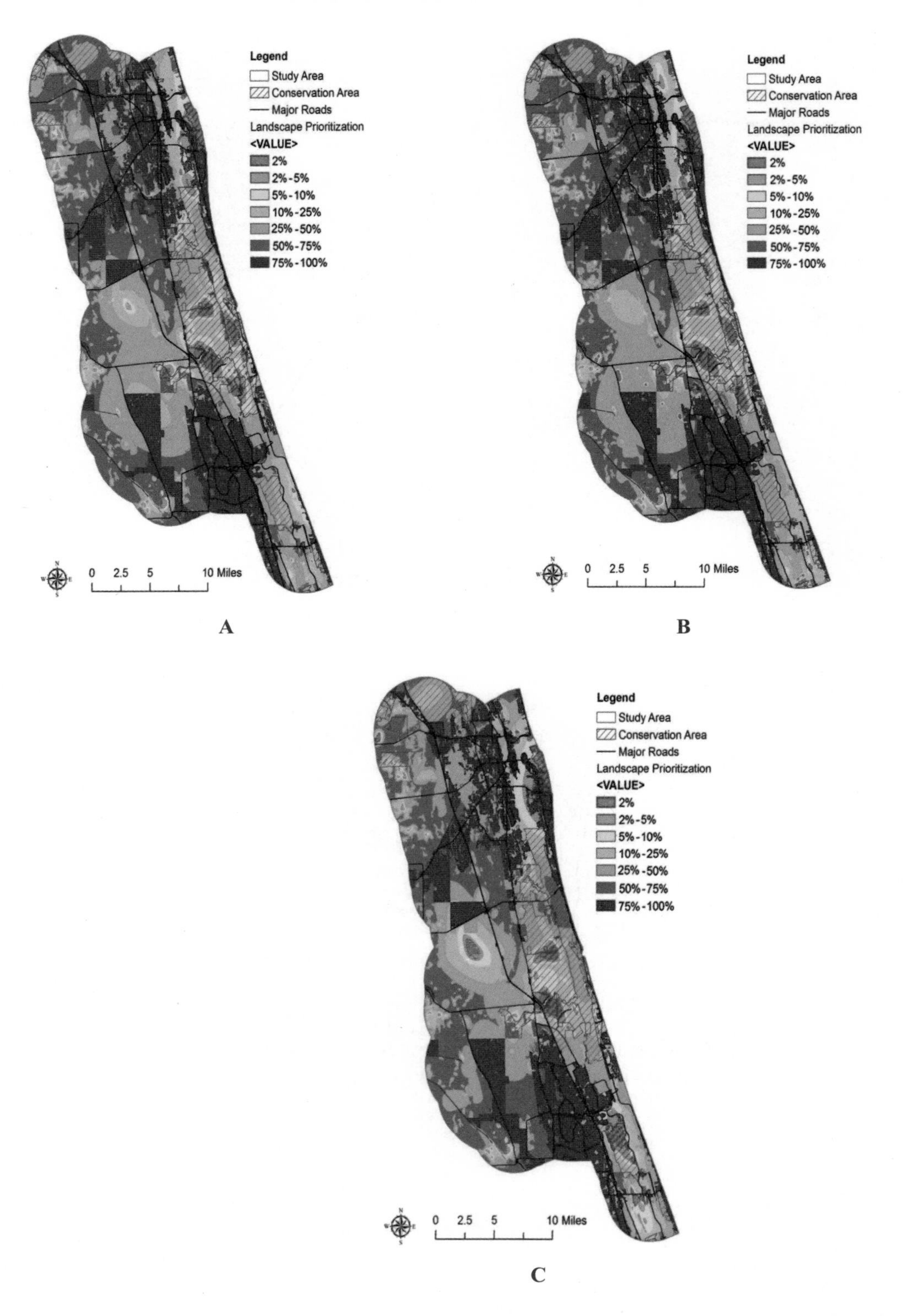

Figure 2-13 Potential areas need to be protected to facilitate biodiversity adaptation to SLR around GTMNERR under (A) 0.5 m SLR, (B) 1.0 m SLR and (C) 2.5 m SLR

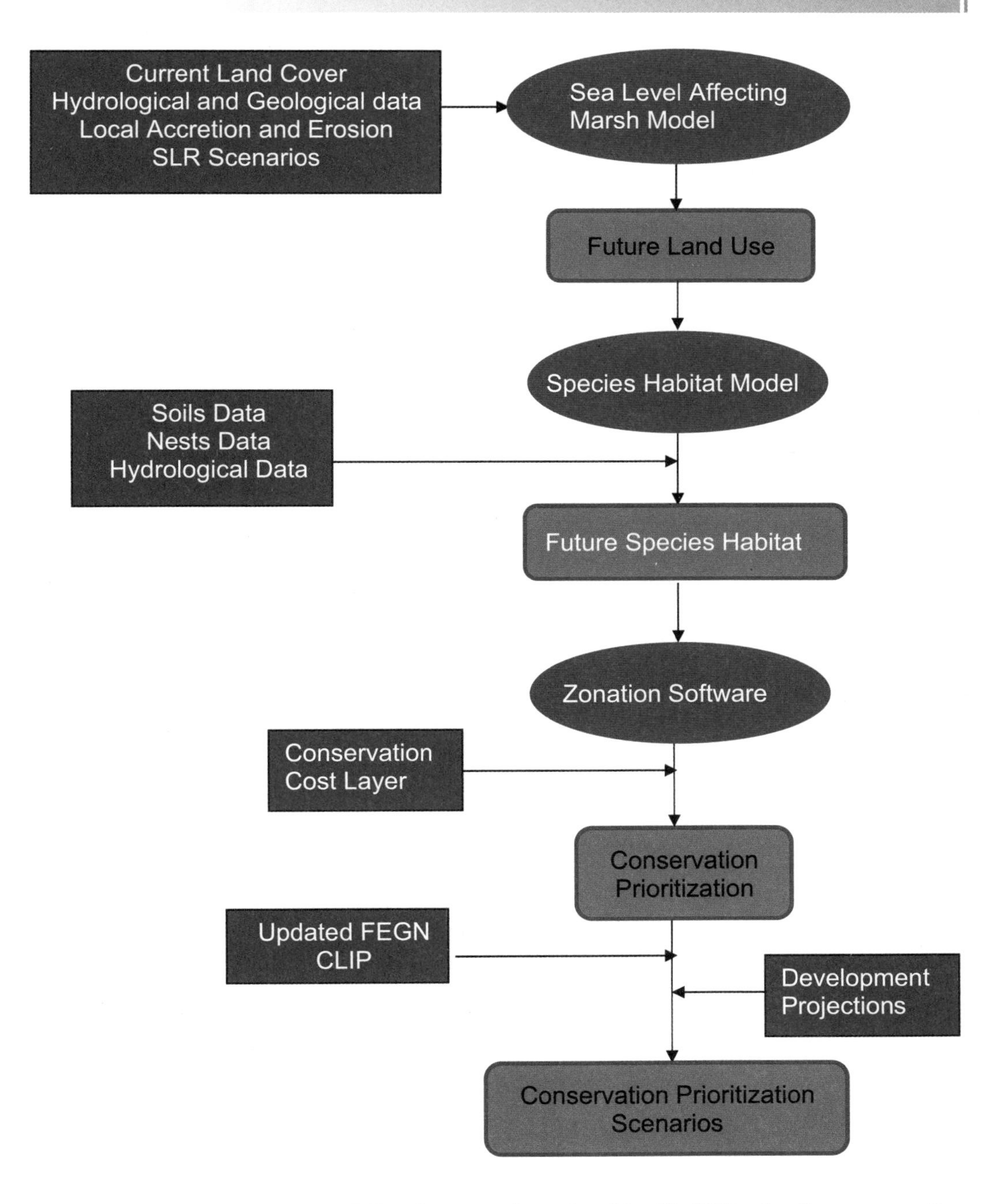

Figure 3-1 The integrated modeling framework used to identify SLR adaptive conservation priorities when including conservation costs in the analysis

The blue oval components are the three models used in the planning process. The purple rectangle components are data inputs and the green components are primary outputs of each analysis

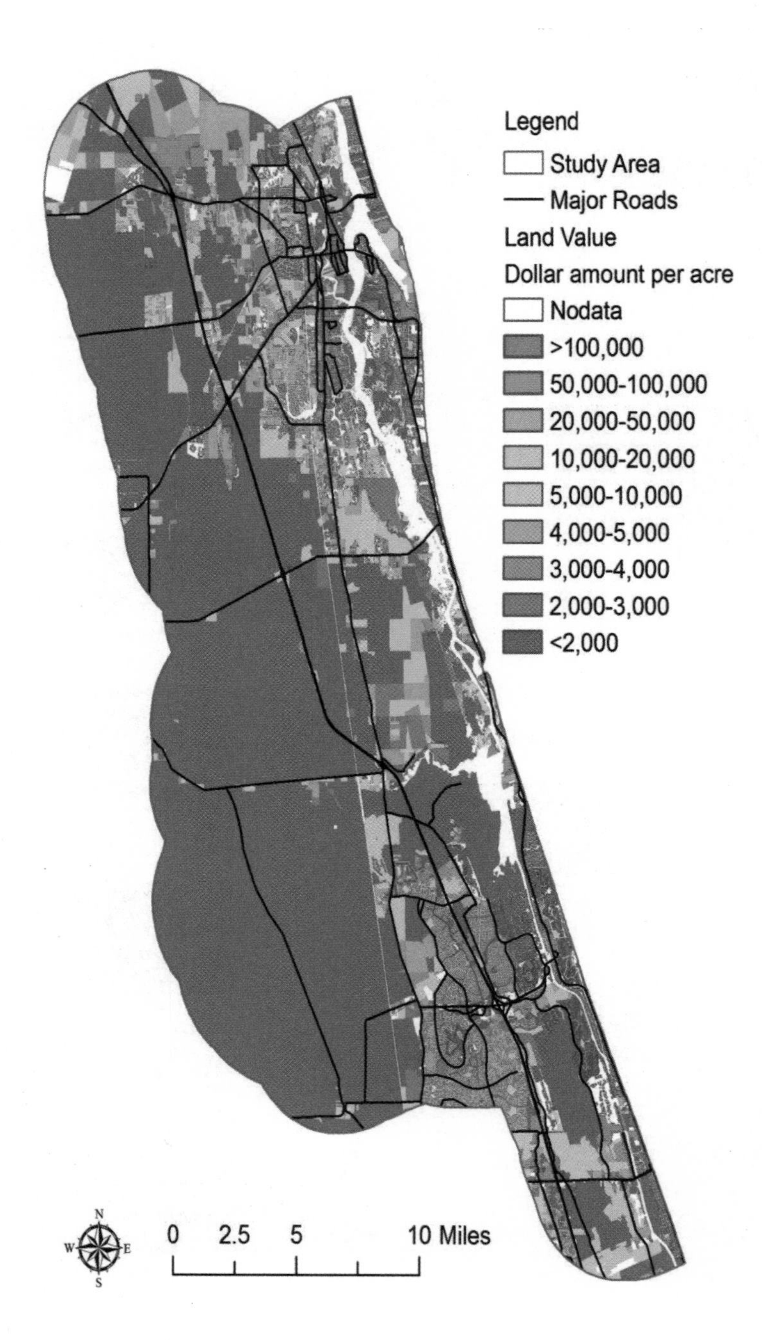

Figure 3-2 Land acquisition costs in the study area

Red color represents areas with highest acquisition cost and the dark green color represents areas with lowest acquisition cost, which are present values in U.S. $ per acre

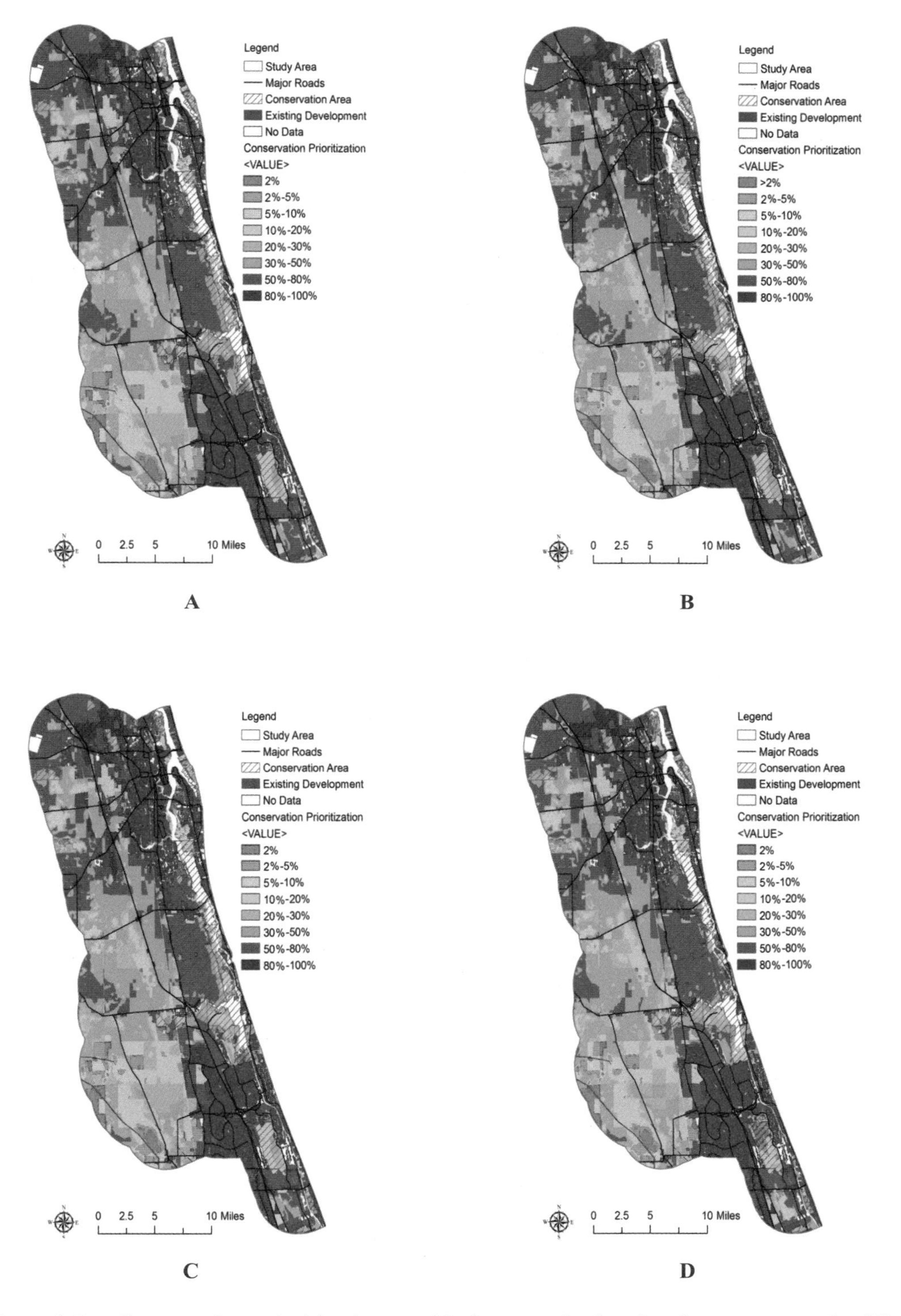

Figure 3-3 Conservation prioritization considering costs in the planning process under (A) current conditions, (B) 0.5 m SLR, (C) 1.0 m SLR and (D) 2.5 m SLR
The red color represents the top conservation priorities

Figure 3-4 Comparison of top 10% conservation priorities under current conditions, 0.5 m, 1.0 m and 2.5 m SLR scenarios

The green color represents top 10% conservation priorities in all SLR scenarios

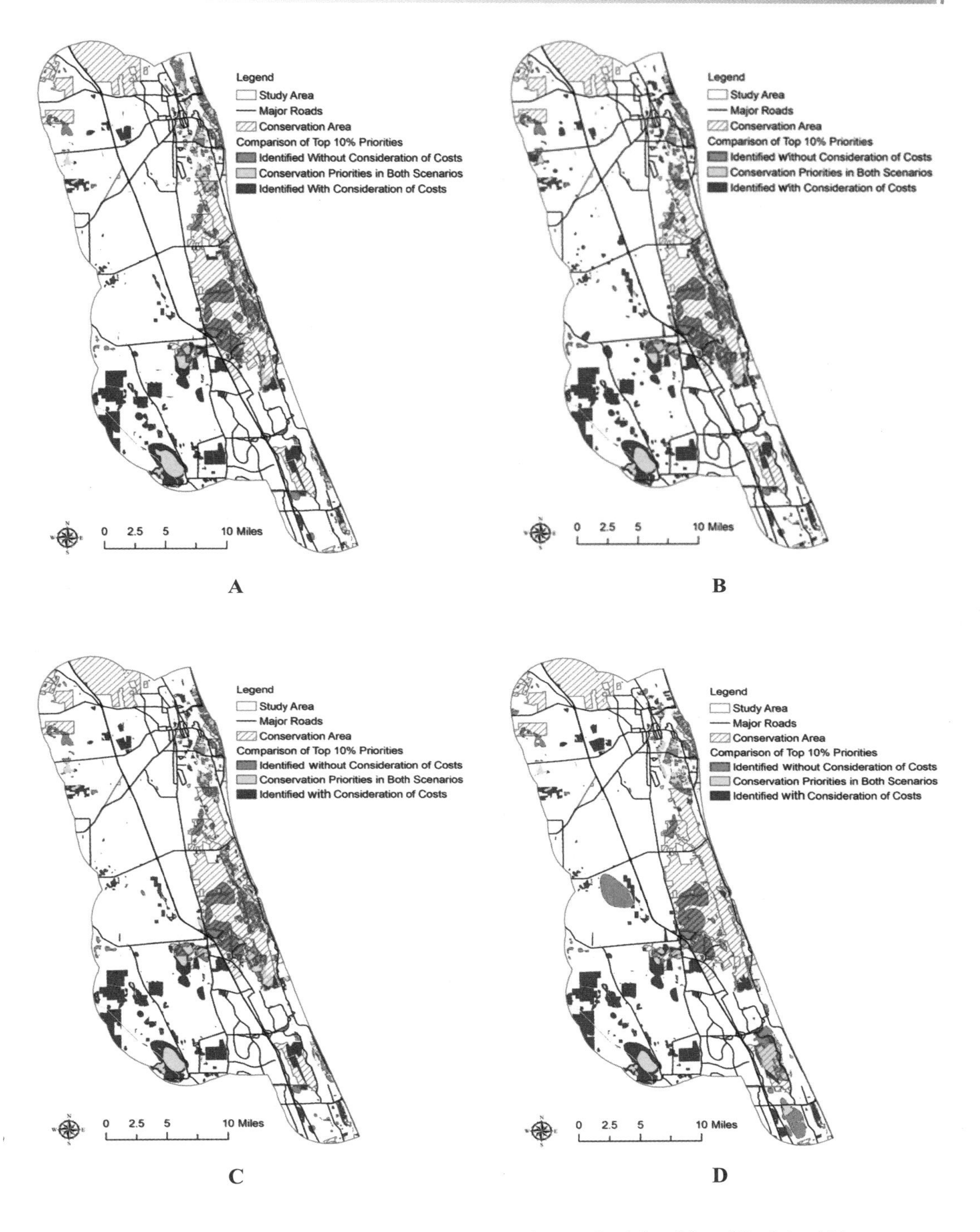

Figure 3-5 Comparison of top 10% conservation priorities identified in (A) current conditions, (B) 0.5 m SLR, (C) 1.0 m SLR and (D) 2.5 m SLR scenarios with/without consideration of economic cost

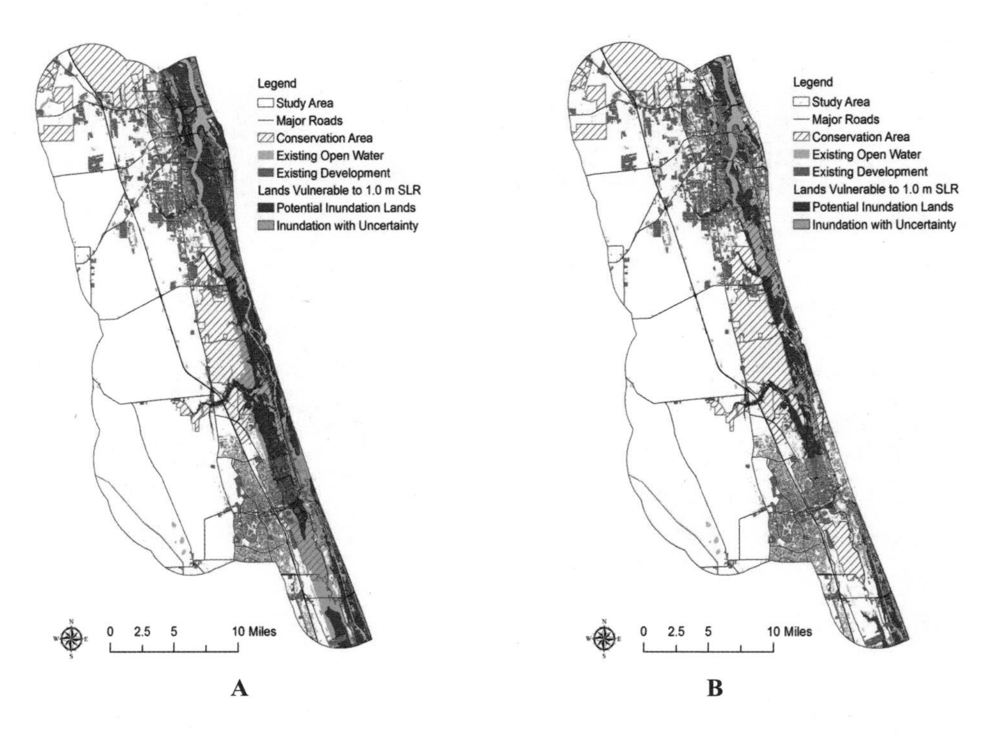

Figure 4-2 Potential inundation areas from 1.0 m SLR based on (A) National Elevation Dataset, (B) Lidar-derived DEM

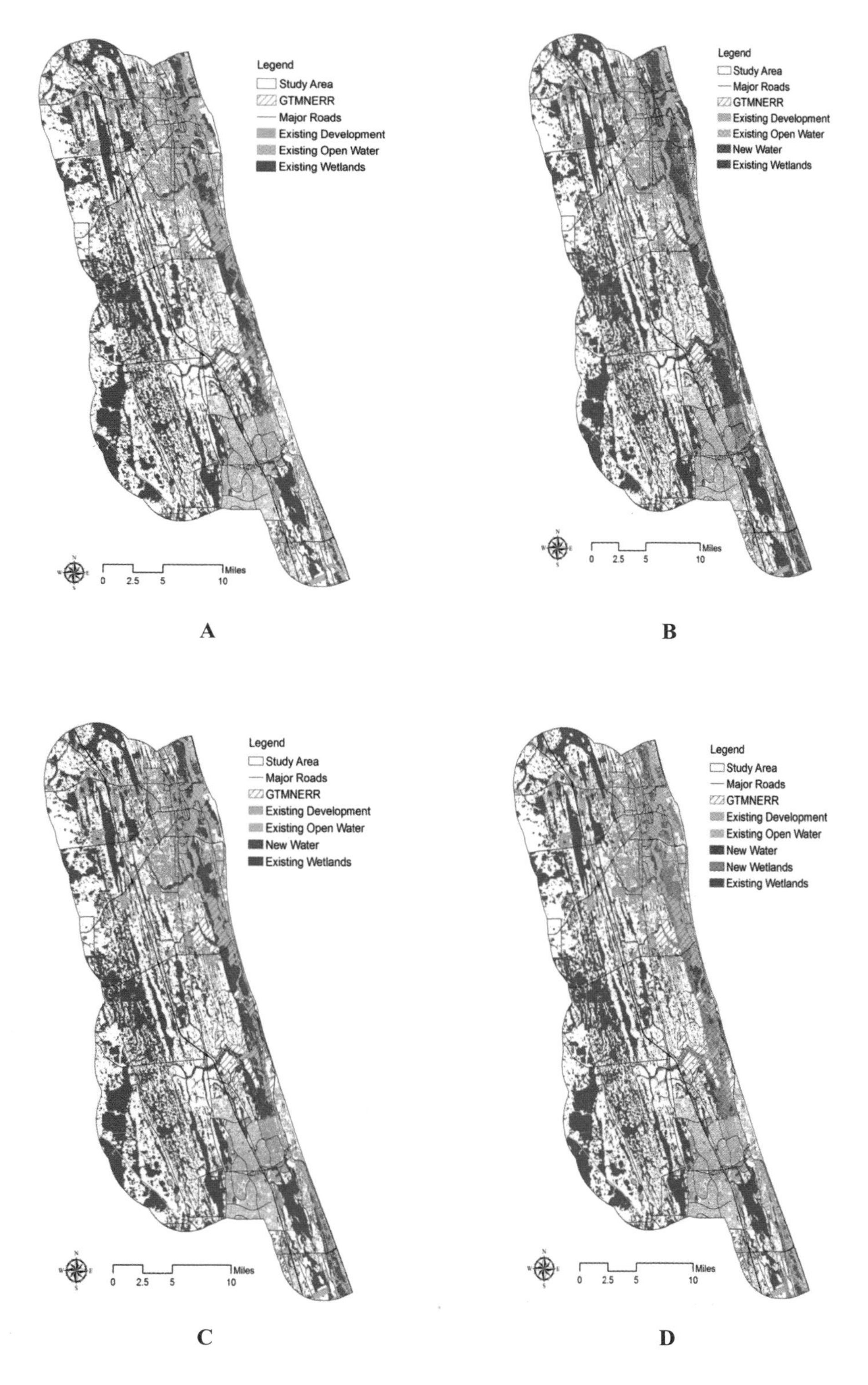

Figure 4-3 Water and wetlands distribution under (A) current conditions, (B) bathtub-NED combination, (C) bathtub-Lidar combination and (D) SLAMM-Lidar combination

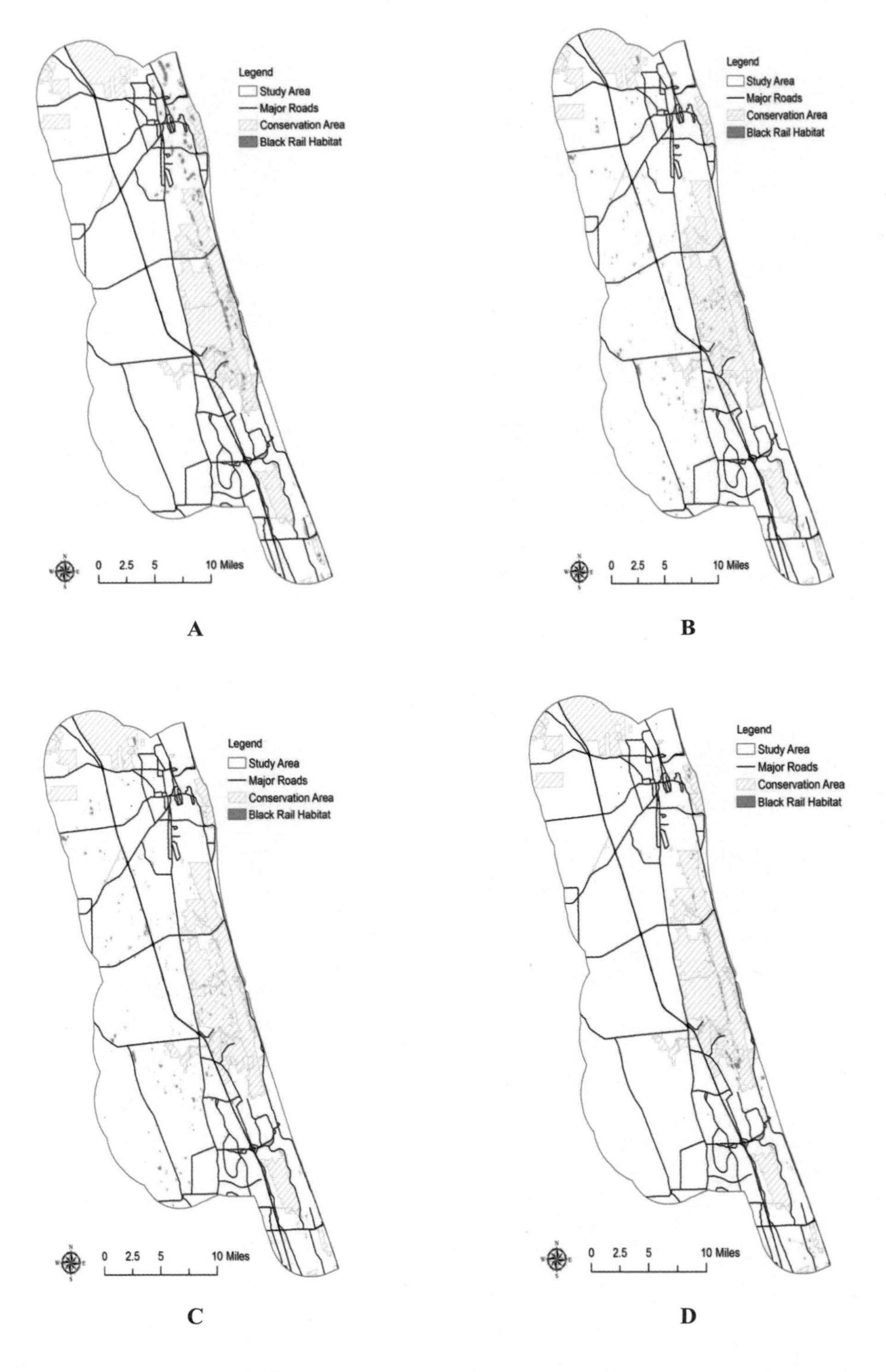

Figure 4-4 Black rail habitat under (A) current conditions, (B) bathtub-NED combination, (C) bathtub-Lidar combination and (D) SLAMM-Lidar combination

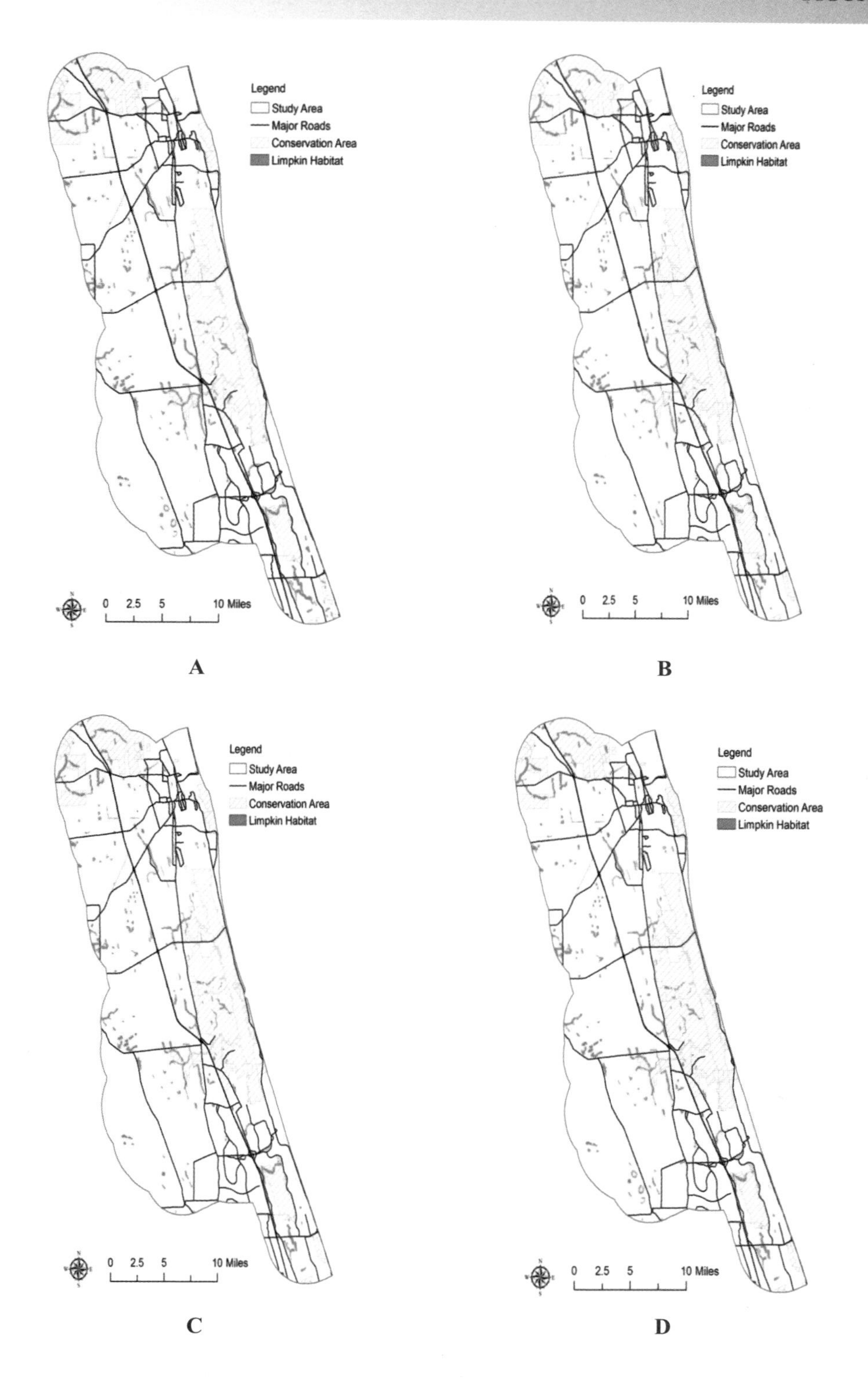

Figure 4-5 Limpkin habitat under (A) current conditions, (B) bathtub-NED combination, (C) bathtub-Lidar combination and (D) SLAMM-Lidar combination

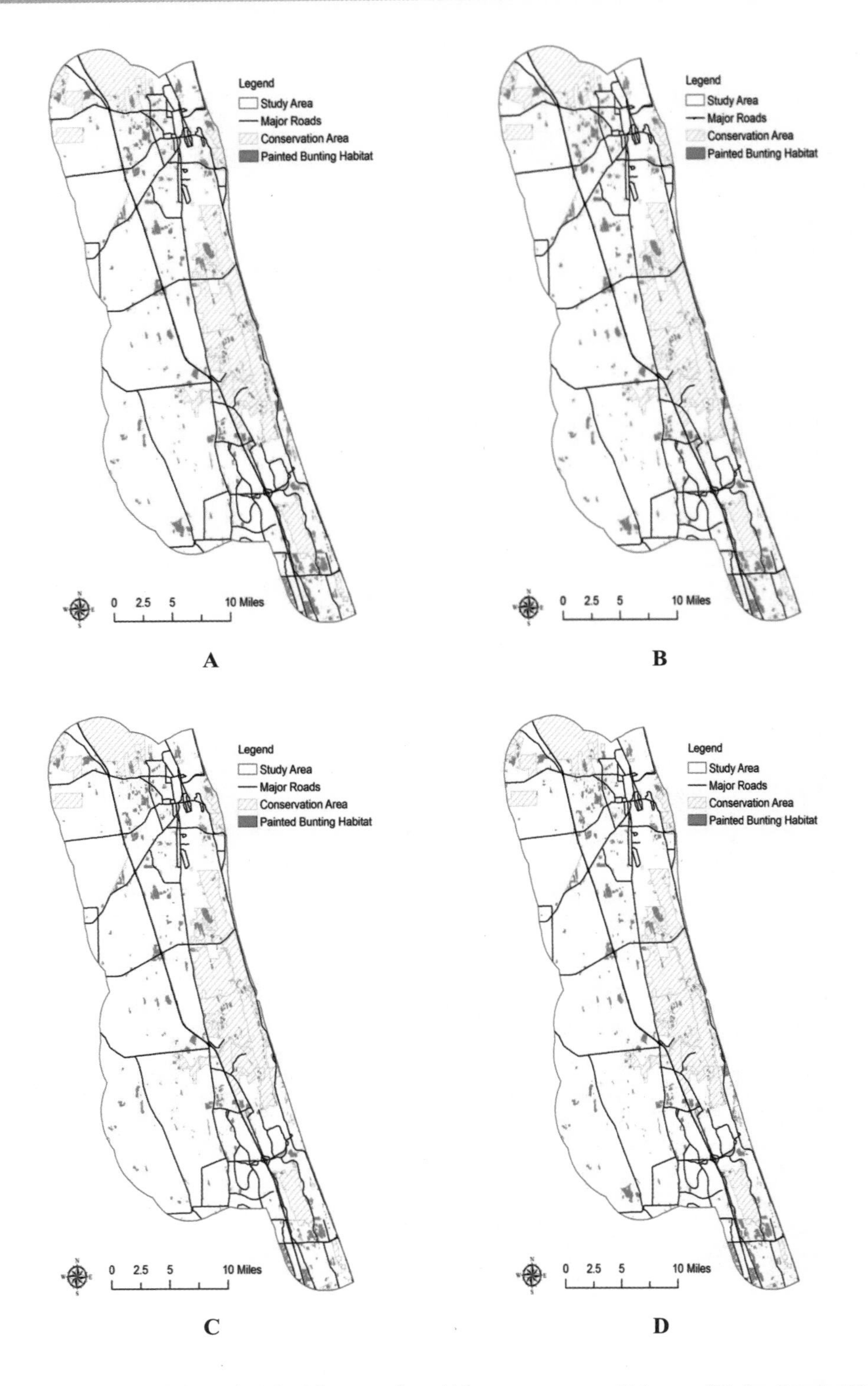

Figure 4-6 Painted bunting habitat under (A) current conditions, (B) bathtub-NED combination, (C) bathtub-Lidar combination and (D) SLAMM-Lidar combination

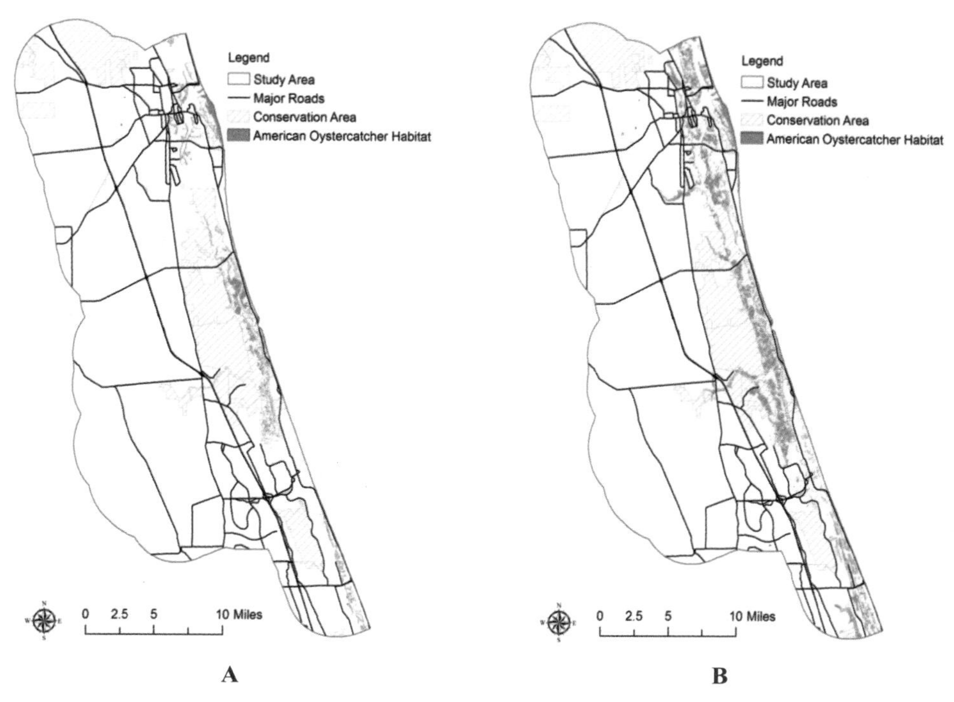

Figure 4-7 American oystercatcher habitat under (A) current conditions, (B) SLAMM-Lidar combination

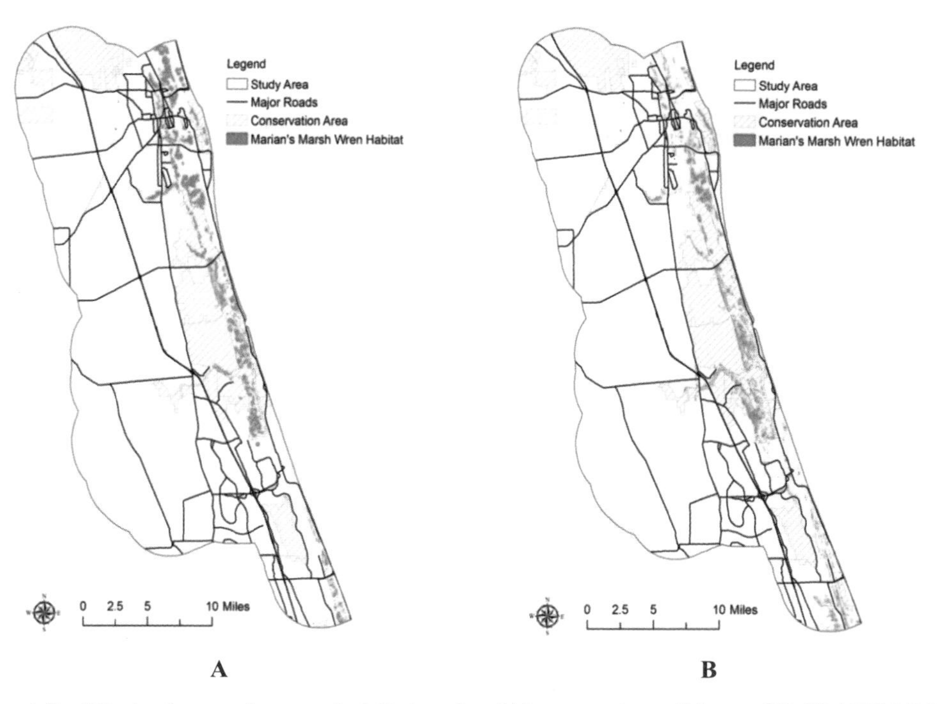

Figure 4-8 Marian's marsh wren habitat under (A) current conditions, (B) SLAMM-Lidar combination

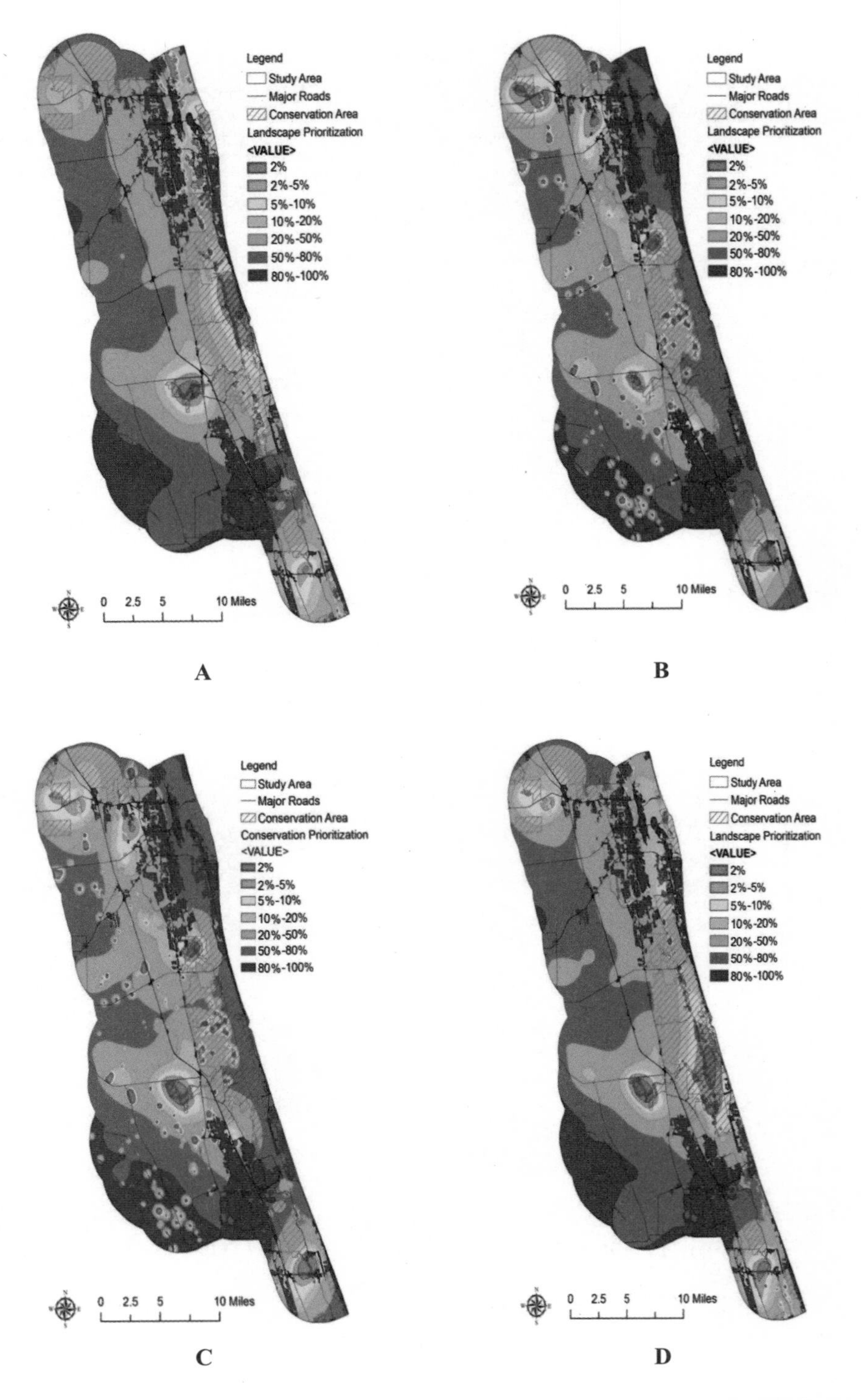

Figure 4-9 Conservation prioritization under (A) current condition (B) Bathtub-NED combination, (C) Bathtub-Lidar combination, and (D) SLAMM-Lidar combination. The red color represents high priorities for conservation

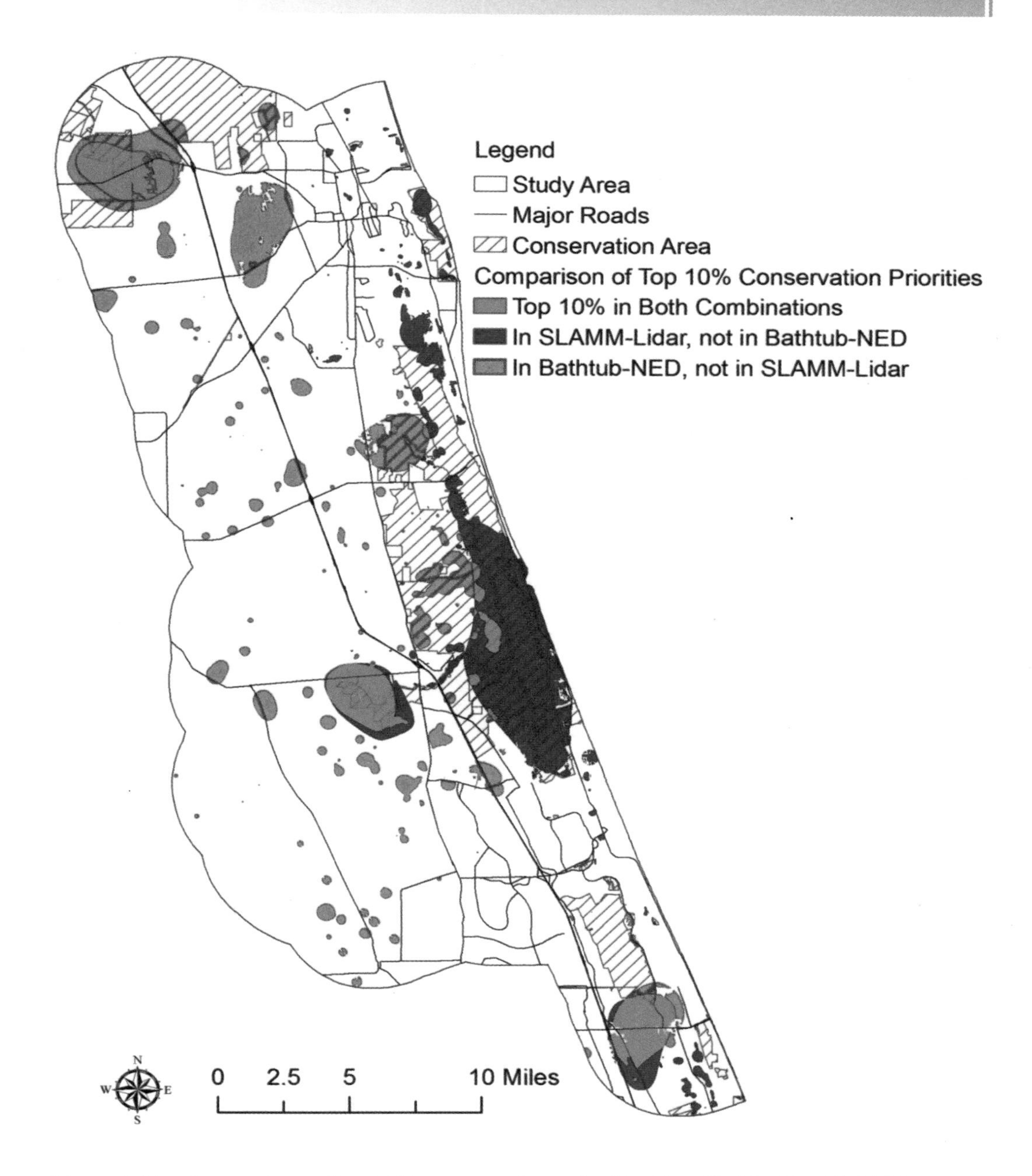

Figure 4-10 Comparison of top 10% conservation priorities identified in SLAMM-Lidar combination and Bathtub-NED combination

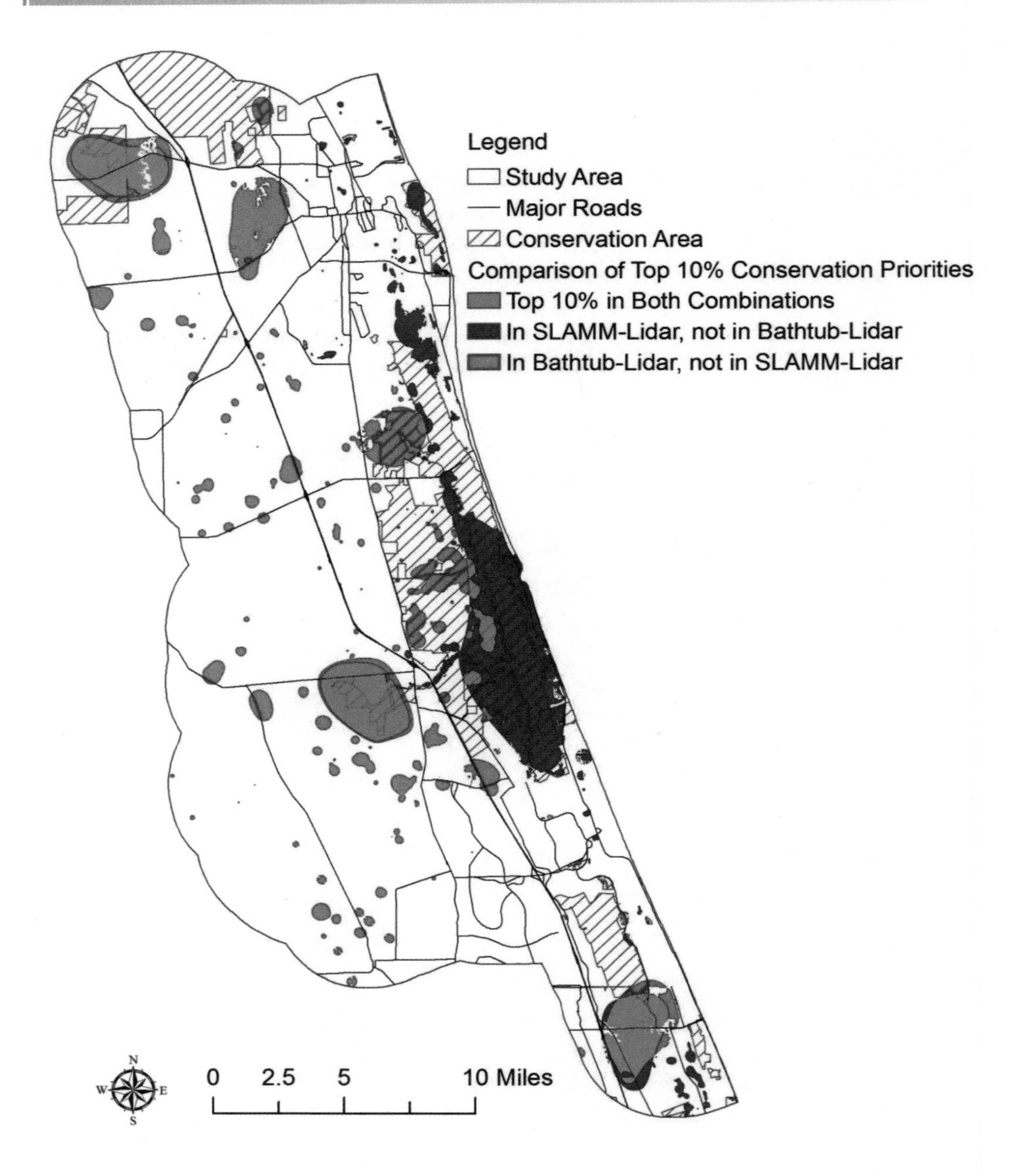

Figure 4-11 Comparison of top 10% conservation priorities identified in SLAMM-Lidar combination and Bathtub-Lidar combination

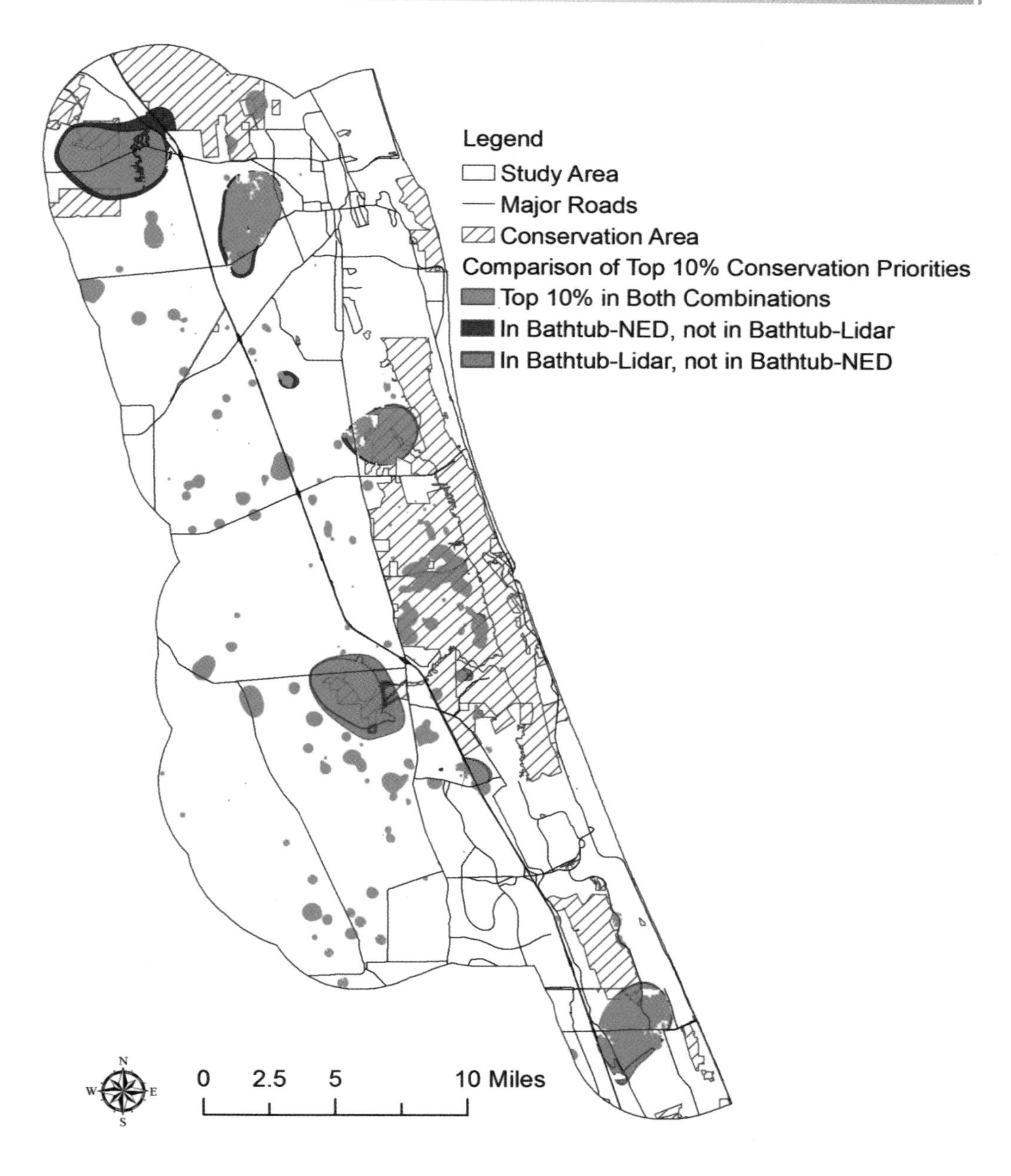

Figure 4-12 Comparison of top 10% conservation priorities identified in Bathtub-NED combination and Bathtub-Lidar combination